EVOLUTIONARY METAPHYSICS IN A POST-TERRESTRIAL WORLD: A UFO-FRIENDLY PROPOSAL

Kathleen Milliere

ISBN: 9781717055552

CONTENTS

THE PROBLEM AND PROPOSED SOLUTION

In this work, it will be argued that there are no such things as "aliens" in the usual meaning of that term. Rather, there appear to be two distinct categories of intelligent non-human beings present on our earth, one of which is fully terrestrial, and the other, not a true "being" at all, but rather a manufactured "bio-machine" which derives from another dimension.

The first of these categories, the "terrestrial alien," consists of various species of biological entities which are alien only in the sense that they are not human. And although it is also the case that they are not, for the most part, *believed in* by humans, they are by no means unknown to us. Terrestrial aliens are familiar to all of us in the form of "monsters," or universal "creature types" common to mythologies in all human cultures since pre-history; a new perspective requires only an ontological upgrade of sorts, which elevates them from the status of "fictitious," to "real." In other words, these so-called "aliens" are not true aliens at all, but rather terrestrial species whose existence on our planet predates that of humans, and has been concealed from contemporary humans by a succession of complicit governments operating under a cloak of secrecy.

The second second type of alien, which will be the primary focus of this work, is truly alien, in both composition and place of origin. And although the two alien "types" are frequently presented in the literature as co-operating in various programs, they differ in fundamental ways.

Specifically, the "creature" aliens, which present as scary monsters, are clearly flesh-and-blood, organic entities, for all of their ferocity, while the other-worldly types stand out in stark contrast as the "little green men" of science fiction, who are distinguished by their high-tech, metallic construction, and emotion-free, robotic deportment.

The "little green men" alien types, who are not native to our planet, come here from outer space, and sometimes crash land on earth and die. But once here, those which survive collect flora and fauna, eviscerate farm animals, and abduct and perform terrifying experiments upon humans, exhibiting an intelligence and possessing technology which far surpasses that of ours. These "other-worldly" aliens are also distinguished by their ability to defy the laws of physics, and it is this feature which will be addressed in the present work.

The work begins with a critical evaluation of some of the key claims found in diverse UFO literature, dividing it into two competing camps:

1. The old-school, "Pro-Science" perspective, which scoffs at the notion that *anything* can be beyond science, and;

2. The "New-Paradigm," occult view, which categorizes extra-terrestrial phenomena as inherently supernatural, and the scientific world view as both unnecessarily limited, and obsolete.

It is concluded that while numerous points from each camp are well taken, neither is completely logical. The present scientific paradigm, it is agreed, *is* obsolete, and many of its practitioners, biased and epistemologically bankrupt.

But it does not necessarily follow from this that the scientific method is *itself* faulty. It might be that it is only the presuppositions, or metaphysical foundations, of the present scientific paradigm which are at fault, because it is these, universally-accepted, points of departure which are ultimately responsible for limiting the scope, and therefore the ability of science to explain a given phenomenon.

In other words, if the underlying metaphysical/ontological world view presently accepted by science expressly rules out things which are being observed, then a scientific explanation for them from within the present scientific world view is indeed impossible, and impossible in principle. But it does not necessarily follow from this that unexplained phenomena at any given time are "beyond science" in an inherent way. Historically, when scientists are presented with unexplained phenomena, they do not automatically conclude that such events are "supernatural." Instead, they assume that science simply lacks the relevant theoretical concepts to explain them at the time. There is no reason to believe that the extra-terrestrial phenomena being observed today differs in kind from historically unexplained phenomena.

Also in accordance with the "pro-science" perspective, it is agreed that the UFOs and ETs observed on earth are real, corporeal, entities. But, contrary to this perspective, it doesn't necessarily follow that the entire cosmos is populated with such beings, even if, as astronauts claim, UFOs have also been observed on neighboring planets like the moon. And, contrary to both camps, the existence of a present-day ET influx does not necessarily support the conclusion that there are unlimited parallel dimensions inhabited by intelligent alien life forms.

Instead, it is conjectured here that most of the so-called "alien" species historically observed on earth and depicted in artifacts dating to pre-human history, are native to our planet. It is further hypothesized that those which *have* arrived here from some other place, all came from the same place – a higher, non-parallel, and physically inaccessible dimension which "contains" our cosmos, and as such, is essentially inseparable from it.

And, finally, it is concluded that none of the other-worldly extra-terrestrial entities which have arrived here from elsewhere, are truly sentient beings. Rather, they would seem to be collectively-programmed, remotely-controlled bio-robots engineered by higher-order entities which are themselves physically unable to enter our lower dimension, but capable of creating micro-tools to observe it indirectly, and alter it and us in limited ways.

Because in the "containment" model proposed, cosmic dimensions are both spatial, and physically inseparable, nuclear explosions on our planet will also affect the higher dimension in which it is contained. So it stands to reason that the "arrival" ETs presently travelling to, and taking up residence on our planet – all of whom began to arrive here immediately after our discovery of nuclear energy – might be coming from that dimension. This would also explain why they all seem to have the same agenda: The dismantling of our nuclear capability.

Their intention, it seems clear, is to do everything in their power to achieve this end, and even though the goal is unquestionably laudable, it doesn't necessarily follow from this that all of their *actions* are ultimately benevolent, or that everything they say is the unvarnished truth. Much of it, especially as it relates to endorsing religion, might be fabricated in order to shame us into following what they have observed are are our *own* moral rules.

Or possibly what they mean when they "confirm" the truth of our religious teachings is that our holy books are true in some *scientific* sense, which, due to deception/censorship on the part of our governments, secular classics scholars cannot accurately interpret. In other words: Christianity as it was in its original, metaphysical, form, before its central concept of a "universal consciousness" was personified into an omnipotent god, and re-packaged for mass consumption.

For example, the propensity of extra-terrestrials to select fundamentalist ministers for the dissemination of information, and oft-repeated ET insistence that the Christian Bible is "true," might be an attempt on their part to convince humanity to reject the mainstream mystical/metaphorical interpretation of this ancient text for a straightforward literal interpretation which substitutes "extra-terrestrials" for "divine beings," and regards geological catastrophes like the Great Flood as natural, as opposed to divinely-ordained, events.

If this is their intent, then this fact would also seem to discredit the contention made by the anti-science ufology set that extra-terrestrials came to earth to provide humans with "spiritual guidance." But what of the claims by anti-science ufologists that the physics-defying powers of extra-terrestrials constitute proof of an occult realm of existence?

It will be argued here that the central, defining, "occult" power possessed by aliens – the ability to move physical objects by "telepathy" – is also used by humans every time our immaterial minds cause our physical bodies to move according to our will. Philosophers have, since the dawn of time, tried and failed to solve the problem of how this works; "telepathy" might be inexplicable, but it is not exclusive to aliens. What *might* be exclusive to aliens is a (non-occult) ability to move physical objects on earth via physical forces which humans are, given our relatively limited perceptual abilities, unable to see.

However one classifies these things, it certainly seems to be the case that extra-terrestrial objects are able to defy the laws of physics as we now know them. But even if this establishes the veracity of claims to what might be referred to as an "occult" realm, the essence of such existence would seem to be more semantical than supernatural, given what we already believe about the comparably anomalous and law-defying behavior of things like quantum entanglement – and (arguably) gravity itself. And it seems similarly unreasonable to assume that the power of ETs is unlimited, or even "unnatural," despite sundry impressive displays of behavior which our current, "material-objects-only" metaphysical ontology cannot accommodate.

It might be that it is the metaphysical ontology underlying the current scientific paradigm, and not the scientific method itself, which is lacking.

With respect to the presently accepted scientific concept of dimensions, it is important to note that the model proposed here differs fundamentally from that of a universe with "parallel" dimensions, which is what is typically meant by the term "alternate dimension." For while the possibility of physically accessible, "side-by-side" dimensions which has been popularized in both science fiction and UFO literature might be tremendously entertaining, from a logical point of view it is both conceptually and empirically incoherent. But even if these "other universes" could be somehow situated in space, the positing of tacked-on parallel dimensions would seem to add little to our *understanding* of either our own cosmos in general, or ET phenomena in particular. At best, their existence would itself require an explanation.

And finally, because the real-physics academic science dedicated to the prevailing concept of dimensions is purely mathematical in nature, and completely bereft of any meaningful conceptual or metaphysical foundations, it is impossible for the non-specialist to comprehend, let alone criticize the work which postulates these multi-universe constructs.

The result is that the present purported "extra dimensions" – which are dismissed as irrelevant by a large and growing segment of the scientific establishment – carry with them certain unfortunate "sci-fi" connotations. Moreover, they would appear to serve no real purpose – apart, perhaps, from further complicating an already untenable general scientific world view by piling even *more* seemingly inexplicable layers of independently functioning systems onto our obviously interactive but allegedly "random" cosmos.

Parallel dimensions were not advanced to explain any newly-discovered empirical phenomenon, whatever they may accomplish from a purely theoretical point of view. But even assuming their existence, they do not shed much light on the present-day influx of extra-terrestrials. Why, if these extra dimensions have been there all along, inhabited by alien beings with easy access to earth, did the extra-dimensional beings suddenly decide to drop in now?

They have come, it will be suggested here, for the sole purpose of eliminating our nuclear capability. Given the difficulties inherent in any explanation which would have ETs travelling to our earth from outer space, it seems that they *must* be coming from an alternate dimension of some sort; the only question is *"What* sort?".

It will be argued here that although their place of origin is indeed another dimension, it is not a "parallel" dimension in either the mathematical, or the popular sense.

In the "containment" model of dimensions proposed here, there are infinitely many dimensions, all of which are situated within a single, evolving universe – not side-by-side in some space-defying "parallel" fashion, but one inside the other, like Russian lacquer "boxes within boxes."

Because all of these dimensions are physically – if somewhat remotely – connected, although the entities which inhabit higher dimensions might not be affected by minor, low-tech explosions and old-fashioned wars down here, a nuclear blast would register significantly, and the all-out nuclear destruction of our planet might prove fatal to higher-order existence as well.

If it is the case that our cosmos is "contained within" theirs, higher-dimension entities would not be able to destroy humans completely without critically injuring themselves; their only real recourse would be to somehow persuade us to abandon our nukes of our own volition.

So the tools that they send down here to attempt to persuade us by flaunting their "superior powers" would be limited to things like threats, terror, isolated acts of violence, and the temporary dismantling of our weapons, communications, and/or electrical systems.

And although these powers might appear, given our dimensionally-limited perceptions, to be "supernatural," the occult and law-defying elements associated with UFOs and extra-terrestrial entities, it is suggested, might be explained scientifically with the adoption of new metaphysical presuppositions and a new theory of evolution.

And lastly, what of the anti-evolution ufological contention that humans were created by hybridizing a terrestrial animal (i.e., "Bigfoot") with super-intelligent extra-terrestials, and that all of the knowledge associated with advanced societies in the past could *only* have come from extra-terrestrial sources?

It will be argued here that this theory, while intriguing, ignores the fact that evolution did not begin with man, but with the Big Bang, and that the course of cosmic evolution is all of a piece; it is not limited to the evolution of humans, or even to the evolution of organic matter. It is the problem of how organic matter somehow "arose" out of inert chemical compounds which is the central problem of evolution, and ufologists who address only the alleged leap from apes to humans are – albeit understandably – driven more by anthropomorphism than by logic.

Before we can even begin to make conjectures about how homo sapiens got their intelligence, we need to find some way of reconciling the laws of physics, which stipulate that disorder increases in evolutionary time, with the increasing complexity of evolving biological systems. And more importantly, to explain how inter-active, *living* systems can emerge from non-organic, inherently independent material matter, which is governed by strictly mechanical processes.

It will be suggested here that inverting this problem provides an answer: If the universe is itself an organic, evolving, multi-dimensional system, and inert physical matter simply a (not-very active) component of the system, then the problem of how life springs from matter does not even arise. The organic universe, multi-dimensional model of evolution presented here is consistent with both the standard laws of physics, and the systems-based, interactive/functional structure of biological laws. Because it also eliminates the randomness required of present evolutionary theory, it can accommodate the possibility that humans were created by hybridizing terrestrial animals with highly intelligent, extra-terrestrial life forms. The contention that advanced

13

technologies associated with ancient civilizations could "only" have come from extra-terrestrial visitors, on the other hand, is debatable.

Following the present "natural selection" model of evolution, most analyses of scientific progress presuppose that humans have been in existence for a relatively short span of time, and that their intellectual evolution unfolded in the same slow, gradual way that their physical evolution did. But both of these entrenched "facts" have recently been called into serious question, and it is now generally accepted in sociological circles that scientific progress is not gradual and accumulative, but accomplished in great leaps and bounds, with one important discovery leading to a multiplicity of derivative new theories and practical applications.

It will be argued here that the theory of natural selection itself is essentially false in every respect, and its companion geological concept, "gradualism," invalidated by all available empirical and historical evidence. Like scientific progress, geological change happens quickly, and frequently catastrophically, especially when the causes of change derive from space. It does not, however, necessarily follow from the fact that ancient civilizations *could* have acquired their knowledge in the usual, abrupt, "spark of inspiration" human way, that they could not also have received at least some of it from extra-terrestrials. How much, it is impossible to say. But it does seem evident that ancient structures and artifacts were profoundly influenced by early human contact with entities from other worlds, and there is no reason to suppose that ancient science and technology were not similarly influenced.

The possibility of alien-acquired knowledge is an intriguing one, and if it is in fact true, as many ufologists claim, that the bulk of our latest highly advanced technology came from reverse-engineering of alien craft, such a fact would lend credence to the claim, even if ETs are not, historically, our regular source of information. And yet, it also seems reasonable to assume that alien visitors instilled the same terror in early humans that they do in us today. In fact, given the remarkable parallels in our respective situations, it is entirely possible that advanced ancient civilizations were visited by alien forms identical to those which are now terrifying us – and for precisely the same reason.

That is to say, it is entirely possible that the relationship between extra-terrestrials and the ancients is exactly the opposite of what is presently assumed, and that rather than *acquiring* their high-tech knowledge from ETs, advanced ancient civilizations were in fact *destroyed* by ETs for discovering – and perhaps even using – the same sort of nuclear energy that has gotten us into so much trouble today.

INTRODUCTION

The following proposal was originally advanced as an alternate perspective on cosmic evolution, simply, but because its central concept, the idea of simultaneous existence across dimensions, nicely accommodates a multi-dimensional perspective on the origin of UFOs, it can also be applied as a proposal for approaching the UFO problem – any solution to which would seem to require not just a collection of credible empirical evidence, but a radically new metaphysical perspective to explain it.

Keywords: *Dimensions, Dualism, Causation, Ontology, Telepathy, Physical Laws, Metaphysics*

Much of the UFO literature to date reads more like a political thriller/manifesto of rights mash-up than a proper scientific study, with ufologists constructing meticulously-researched papers compiling evidence purporting to establish the existence of both an extra-terrestrial presence on earth, and a military conspiracy to withhold any information about its existence from the public – combined with a declaration of our right to know and a demand that scientists treat the subject in a timely and respectful manner.

Added to this are complaints about an arrogant, politically-biased press which simply cannot resist the opportunity to sell a good laugh at the expense of anyone brave and civic-minded enough to submit a report, and a patronizing, secretive military which reserves for itself not only the ultimate right to rule the universe, but the ultimate say on who is permitted to understand it. And while it is sometimes entertaining, none of this is especially useful, as none of the groups being criticized can reasonably be expected to alter their respective agendas.

The military's first duty is neither to the people, nor to the government; its responsibility is to protect the state, and, practically speaking, there is no point in attempting to "command" it to do anything. The military is in charge of national security and if it deems information to be potentially harmful to the public, then it will classify that information. They are the experts in this matter and we have to trust them to do what is in our best interest. Also we have no choice.

As for scientists, it simply is not reasonable to attempt to impose theoretical demands on people who have acquired a particular body of knowledge, and are trained specifically to apply the rules contained therein. If the existence of some phenomenon which violates these laws is alleged, they will either ignore it, or they will attempt to explain it in a way which *does* agree with the rules that they have learned. What else can they possibly do? They can rightly be called closed-minded, illogical, unethical, or all of the above, but they cannot be forced to countenance the existence of objects, entities, or behaviors which violate the only principles of knowledge at their disposal.

In the case of UFO's, they are being asked to verify phenomena which also violate the most fundamental of principles upon which these beliefs are based: Single-substance materialism, biological individualism, and even causation itself. And to make things worse, these concepts are simply *presupposed* by science; they are not understood by scientists or addressed in any of their texts. They represent metaphysical and epistemological concepts which fall so far outside of the scientist's domain that even the rare practitioner who *is* willing to tackle them typically possesses neither the education, nor the conceptual skills required to do so. J. Allen Hynek probably said it best when he despaired of any solution given the present state of science and cried, "Is there a philosopher in the house?"[1]

The present approach to the UFO problem, then, which essentially amounts to political demands that the government release classified historical information which may or may not exist, so that unspecified non-scientists can independently "test" the veracity of past incidents which were not recorded and treated with the requisite scientific scrutiny and respect in the first place, has not proved especially useful. But even assuming that this secret evidence really does exist, and that the government can be persuaded to release it, what then? We would simply have at our disposal a collection of historical reports of UFOS. But we already have that. The problem is what to *do* with it. Rather than take Hynek's approach[2]

---

[1]   "The Edge of Reality: A Progress Report on Unidentified Flying Objects; 1975.

[2]   J. Allen Hynek (1910-1986) was an astronomer harboring extreme anti-UFO beliefs. Indeed, he ridiculed the concept, especially when it came to the "little green men" part – which is why he was invited to serve as an "independent" consultant on

to UFOs, which is to attempt to devise a rigorously conservative, statistics-based test for verifying the existence of something for which we have no concrete, physical evidence, the proposal here is to begin by assuming the approximate truth of the various reports made to date – hundreds of thousands of UFO reporters can't *all* be liars, after all – acquire in this way a general idea of what we are dealing with, and then attempt to construct an alternate scientific/metaphysical perspective in which the presence of such phenomena would be not only permitted, but positively *expected*.

The problem with applying an evidence-based approach in the case of UFOs is not just that, properly applied, this

---

the government's "Project Blue Book."

Blue Book, which has recently been de-classified, was not a scientific investigation into anything. It was designed specifically to function as a public relations exercise, dedicated to debunking all things extra-terrestrial – while military scientists continued to conduct their classified research into the subject.

After examining the evidence, Hynek emerged as a no-holes-barred believer and scathing critic of both the approach and the conclusions Blue Book contained, and turned his wit to government scientists' decided disadvantage.

Hynek's own books advocate a philosophical approach to the problem, and would be considered required reading for anyone seriously interested in UFOs, although many of the epistemological concepts that he introduces might be unintelligible to readers without at least some background in the philosophy of science.

Dr. Hynek was also something of a rock star in the UFO field, critically assessing evidence on a case-by-case basis, giving public lectures, and even appearing in Steven Spielberg's "Close Encounters of the Third Kind."

would essentially eliminate all of the evidence – virtually none of which would appear to meet even minimum requirements for serious scientific consideration – but that since any scientific paradigm automatically rules out the existence of phenomena which violate its laws, scientists subscribing to the current scientific world view would not be able to explain UFOs scientifically even if they did acknowledge their existence.

An alternate scientific world-view in which the sorts of events associated with UFOs are not only consistent, but expected, would eventually enable scientists to formulate a new set of laws which would be capable of describing the phenomenon in the standard, hypothesis-worthy way, instead of just dismissing UFO reports as the products of over-active imaginations.

It would also relieve individuals interested in the subject of the burden of acquiring impossible-to-acquire historical evidence in order to make their case. For if the presence of UFOs is both as real, and as pervasive as past reports seem to indicate, then there is no reason to suppose that the phenomenon is somehow "over," and that past reports represent our only way of obtaining evidence. In fact, the opposite is true.

If, as it is generally assumed, perception is at least in some measure determined by theory, the acceptance of a full-fledged scientific world-view in which UFOs are to be expected will increase our ability to perceive them correctly when we encounter them, rather than simply dismiss or attempt to explain away such perceptions. And, of course, eliminating the ridicule which invariably accompanies any report in our present paradigm will make it desirable for people who do perceive them to come forward with reports.[3]

As for those brave individuals who have already come forward,[4] their evidence, while perhaps not scientifically verifiable, certainly tends to reinforce both the reality of the phenomenon, and the need for a fundamentally different world-view which can accommodate it. Logically speaking, there is no reason *not* to believe them.

The first thing to note about UFO reporters is that virtually all of them either did not believe in, or had no opinion about UFOs before witnessing an event. According to some statistics, only about one in 10 experiences is ever reported, and people who do make a report do so at great risk of being ridiculed, humiliated, and worse. Yet their experiences are so compelling that they feel an absolute duty to inform *someone*, and they cannot be persuaded that they are "mistaken" about what they have seen.

But most telling, rather than being "imaginative," these reports have a repetitive, positively tedious sameness

---

[3] See Appendix A for a summary of some of the "whistle blower" reports from the original astronauts, virtually all of whom claimed to have seen UFOs.

[4] Many UFO reporters have risked far more than just ridicule, as retired Air Force pilots – dedicated, patriotic servicemen who are effectively committing treason by such disclosures – are beginning to connect with each other online, and even going public with what they know, simply because they believe that they have a higher duty, and that it is in the best interest of their country for the public to be aware of the problem.

about them; reading collections of such reports for even a short time inevitably results in both "desensitization," and the realization that it is probably pointless to continue – they are all alike and no new information will be forthcoming.

UFO reports remain interesting in a conceptual sort of way, taken in their entirety, but because they are typically advanced by the sorts of individuals not usually given to poetic expression – policemen, pilots, farmers, and the like – they tend not to be terribly interesting as individual stories. But because it is impossible for a rational person to dismiss *all* of these stories as combinations of hallucinations, hysteria, and swamp gas, the problem is not so much a matter of "validating" them as it is one of *explaining* them. Here is a short list of some of the common elements of the various credible UFO reports on record, and the theoretical beliefs outlined in the ufology literature:

1) UFOs travel at physically "impossible" speeds, and behave in ways which violate many other physical laws as well.

2) UFOs "vanish" into thin air when they are observed.

3) UFO craft and lights appear to be "alive."

4) UFO craft are operated via "mind control."

5) Extra-terrestrials communicate via telepathy.

6) Extra-terrestrials have influenced our evolution/biology.

7) Most extra-terrestrials pose no threat to national security; indeed, they are both non-violent and intensely interested in our well-being.

8) Extra-terrestrials communicate the desire to help us, and attempt to teach us to stop destroying our planet.

9) Extra-terrestrials try to persuade us to stop being so destructive, but they refuse to interfere with our freedom to do as we please.

10) Extra-terrestrials often carry overtly religious messages.

11) Extra-terrestrials do everything in their power to avoid detection, and try to select isolated, sparsely-populated locations in which to land their craft.

12) Extra-terrestrials tend to approach only select individuals when they have information to convey, and always avoid large groups of people.

13) Malevolent extra-terrestrials abduct and perform intrusive medical experiments on humans, extracting tissue and bodily fluids.

14) Extra-terrestrials can interfere with our power grids and defense systems, knocking out electricity and de-activating nuclear missiles telepathically.

15) The human/animal hybrids depicted in ancient Sumerian, Egyptian, and indigenous American cultures are not fictitious, but represent real

creatures created by extra-terrestrials for unspecified purposes.

16) Extra-terrestrials from outer space originally created humans by hybridizing their own genetic material with that of lower-order terrestrial animals.

17) Ancient extra-terrestrial visitors to our planet hybridized their own DNA with that of human leaders like King Gilgamesh in order to advance our evolution.

18) Advanced ancient civilizations received their knowledge and technology directly from extra-terrestrial visitors to earth.

19) The blueprints for ancient Sumerian extra-terrestrial technology are contained in artifacts buried in what is now southern Iraq, and the quest to find them is what all of the turmoil in the Middle East is *really* about.

20) Much of our advanced technology has been acquired from extra-terrestrial sources – either directly, through communications between military scientists and visiting extra-terrestrials, or by the reverse-engineering by secret military-affiliated industrial tech companies of recovered ET craft and debris.

21) The American and Russian governments "manufactured" the so-called "Cold War" as a cover for increased military funding, and the "Star Wars" weapons first requested by

President Reagan – who had been visited by, and was terrified of UFOs – were in reality intended to combat hostile extra-terrestrial forces.

22) Extra-terrestrials kill and surgically mutilate cattle, apparently for the purpose of conveying a "message" to alert/help mankind in some way.

Rather than laboriously attempting to dissect independent reports, one-by-one, for the purpose of establishing their veracity, the aim here is to construct a broad overview of the situation, and then use it to create a composite model for describing the general "sort" of entity/ies we are dealing with, including their usual beliefs and behavior.

Once this is done, the task becomes one of attempting to devise scientific explanations for both. It might be well to dispense with the final attribute listed above first, since it is probably both the best-known, and the most inexplicable.

Regarding the abundance of bovine-specific animal mutilations, the simplest explanation would be that aliens are interested in obtaining samples of as much flora and fauna as they possibly can, and the cattle at issue fall under the rubric of "large animals."

If, as frequently happens, a craft lands in an isolated, ranch-worthy location in the American southwest, cows are available. And when one considers the alternative of, say, attempting to catch and mutilate a mountain lion, or an antelope, they would furthermore seem to be the logical choice.

But perhaps this explanation is a little *too* simple, and the ostentatious "gathering of samples" which has so often been observed is simply staged – an otherwise-pointless exercise designed to shield what aliens are *really* up to. This seems much more plausible, as the sample-gathering has been going on for 70 years, which seems more than sufficient for highly-advanced intelligences to complete their rock collections and perform high school biology experiments on the same terrestrial life forms.

So perhaps there is something about the genetic make-up of cows which renders their organs especially suitable for cloning, or their personality traits especially appealing for hybridization. An alien interested in transporting and/or hybridizing terrestrial genetic matter on a large scale would need a readily-available source of material, after all. And what better than a contented cow to counter-act the all-too-often violent nature of other large animals, if one is creating a new species of mammals?

This is not intended to be flippant, but only to point out that it isn't always necessary to attribute symbolic significance to alien behavior which might be otherwise explained in a completely practical sort of way. Contrary to what so many ufologists seem to think, there probably isn't any mysterious and enlightening "message" being conveyed through mutilated cows here; it is much more likely that this business doesn't have anything to do with our best interests at all.[5]

---

[5]   In fact, the unexplained, seemingly "surgical" mutilation of animals is not limited to cattle, but includes all manner of wild animals as well, and is global. In America, where it is a serious

But suppose that these cow-mutilating aliens did have some message to convey. Would they really take so much trouble to make contact simply to play guessing games, especially since if ETs want to communicate with us, they can just use telepathy? Their message seems to be of both a pressing and a serious nature – indeed it is grave – and they certainly do not wish us to misinterpret it. This is connected essentially with the next point, which has to do with the religious nature of the ET message.

Many ETs are purported to espouse religious beliefs, and most science-minded critics interpret this as a red-flag reason to discredit the reporter: We all know that religion is "unscientific," and therefore an alien from a technically-advanced world would never speak like this. So the reporter is summarily dismissed as a religiously-motivated liar.

In cases where the reporter claims that the alien has delivered a message specifically to *himself*, and instructed him to take it to the world, the reporter is further branded as a narcissistic self-aggrandizer with a Messiah complex.

Add to this the fact that these allegedly religious aliens also claim to be from planets like Venus, and tend to present themselves to spiritually-inclined, science-averse humans, and the evidence would seem sufficient to dismiss such reports outright as completely undeserving of our attention.

-----

problem for ranchers, there are frequent reports in the news of cattle mutilation, and the subject is under federal investigation.

This is the standard position taken by science-minded ufologists. But is it really logical to simply *dismiss* – and indeed to ridicule, which is also the "scientific" standard in such cases – these reporters, especially when there are so many of them that religion really does seem to be the usual thing, ET-wise?

Consider, after all, the options available to an extra-terrestrial who wants to make contact with earthlings for the purpose of warning us that our nuclear technology is potentially catastrophic to the larger universe as well as to ourselves. Suppose that his alien race has been observing/monitoring earth for millions of years, and is intimately acquainted with its history, including the lost histories of advanced ancient civilizations – which are available to *us* only as remnants, in the form of artifacts so "other worldly," and texts so fantastic that they would seem to defy literal interpretation. Except, that is, to those who do not reflexively apply the observation-based empiricist agenda which was ushered in at the dawn of the "Enlightenment," and is now entrenched as *the* scientific world view. Because such thinkers also tend to reject the concept of *homo sapien* supremacy for a richer ontology of "what there is," they would naturally be more receptive to the possibility of non-human visitors to earth.

So despite – or, more accurately – precisely *because* the empiricist-based academics tasked with interpreting ancient texts were so blinded by the inherent bias which inevitably accompanies advanced university "indoctrination," they missed what is obvious to ufologists today, who approach the problem from a meta perspective, and a preconceived "belief-in-the-other-worldly" bias of their own.

Ufologists also have a very different, and decidedly non-academic, agenda: Replacing the traditional "mythological" interpretation of events and beings outlined in ancient texts with a literal interpretation not only permits us to make sense of present-day "supernatural" events, but is critical to the future of humans as well as to the larger universe in which earth is contained. This is because if we continue to define extra-terrestrials as "mythological," and suppress/re-define all observational evidence of their existence, we leave ourselves vulnerable to a very real alien threat, however we care to define it in terms of semantics.

To our alien visitors, on the other hand, *we* are the threat, and we must somehow be made to understand why, and persuaded to change our behavior. This would require alien visitors to select as spokesperson a human who would not only listen to them, but would agree to convey their message to the world.

If you were a visitor to earth intent upon conveying an urgent message of an essentially moral nature, would you select a highly-educated individual of any variety, let alone a scientist, for this purpose? The briefest observation of our earth would assure you that someone with advanced academic education would be the last person to listen to you.

It would also reveal that the intelligentsia on earth are appallingly ignorant when it comes to matters of basic science. Could you expect a professor of literature, for example, or an architect, to understand what you meant by saying that you were from a different dimension? But every earthling knows what Venus is, even if they don't know anything *about* the planet.

So the alien from a higher dimension might decide to approach someone unprejudiced by too much education – someone who reads supermarket tabloids, preferably, and is accustomed to stories about aliens – and couch his message in terms that his human listener could easily understand, by claiming that he is from "Venus."   He might correctly suspect that the average human would be far more likely to warm to a sensitive, spiritual, solar system neighbor from beautiful downtown Venus than to, say, a human-machine replica injected with artificial intelligence from an incomprehensibly advanced other dimension. And that, in his excitement about the contact, would be liable to overlook little details like the fact that there are almost certainly no underground civilizations on Venus, and if there were, a Venetian would be unable to breathe our atmosphere.

The science-fiction aficionado might be similarly taken in if the alien claimed to be from a distant planet or galaxy, especially if said contactee were familiar with that particular cosmic location. Logic, after all, does not rule out the possibility of intelligent life elsewhere in the universe; indeed, it would seem to be highly *improbable* that we are the only intelligent species in the cosmos. Here at last, he would conclude, is proof positive that we are not!

Despite its considerable surface appeal, this so-called "scientific" reasoning is faulty because while it might be highly improbable that we are the only intelligent life in the universe, it is even *more* improbable that we, plus some isolated species in a distant galaxy, are the only intelligent life in the universe.

From a logical point of view, the two most probable scenarios are:

1. The universe is teeming with intelligent life forms of every description, on every planet, and we are all aware of at least some of our near neighbors, and;
2. We are alone in the cosmos.

A highly improbable alternative is that we are the only intelligent life form within (x) number of light years, but there are others almost exactly like us – too far away for us to know about. Nevertheless, they know all about us, and can violate the laws of physics to travel here – and do so simply to bring messages of universal peace.

The least likely alternative is that the universe is teeming with diverse and intelligent biological species, near neighbors who know all about us – only we can't see them because they all live underground.

These extra-terrestrial neighbors are organized into hierarchical, religion-based empires which cast Planet Earth out eons ago because humans were too brutish and unenlightened to get along with.

But ETs have now, with our development of nuclear technology, decided to renew contact, make friends, and even participate as active partners in a new global earth government – for both our own good and the good of species on distant, inaccessible planets. (The author is not making these up. They are standard assertions in the literature, albeit highly condensed.)

Given the above set of improbabilities, if, as becomes increasingly evident, a diversity of alien forms has indeed arrived on earth, and cannot reasonably have come from anywhere in our cosmos, the most likely explanation is that they are coming from an alternate dimension.

Presumably, the inhabitants of this dimension — wherever it is — have been observing us all along, and, while they do not find us particularly interesting, are obliged to not only visit, but to take drastic action in cases where our destructive behavior threatens their existence as well. This would explain both their presence in, and the destruction of, advanced civilizations in the past.

Because our dimension is physically inaccessible to them, the visiting entities sent down here are not true biological beings, but have been manufactured – in accordance with carefully-compiled higher-dimension observations about our zest for things like religion and science fiction. As non-biological beings, they can easily "survive" in our alien biosphere.

They can also "speak" all of our languages, and, because they have been programmed to mimic our collective conceptions about aliens and superpowers and spaceships, appear to us to be authentic, technologically-advanced, inter-stellar travellers. Note that even their attire and modes of transport consist in made-to-Hollywood-order fare, complete with tightly-fitting metallic techno pantsuits and jazzy blinking "instrument panels."

Missionary ET, in other words, has presented himself looking precisely as we expect him to look, bearing a message couched in what he has observed to be our planet's fundamental beliefs, for the express purpose of preventing us from critically injuring his *own* higher dimension – in which our planet is unfortunately contained.

Note that although there is believed to have been intermittent, or perhaps constant, monitoring of our earth by aliens all along, most accounts in the literature have all of the different species of aliens currently amassed here arriving at the same time – co-incidentally with the development of nuclear technology.

Most agree that this assortment of alien "races" – which consists in between four and 12 different types, depending on the source – differ from each other rather dramatically in terms of both behavior and appearance.

According to the literature, upon their arrival, all of the assorted different ET species claimed to have come from correspondingly different locations, all of which are too distant for us to know anything much about, except that they are in fact, according to our understanding of astronomy, real places.

The aliens did not indicate whether there were any other species inhabiting their respective planets of origin, but they did refer to their home planets by slightly different, "alien-sounding" names – which posed no problem because the aliens also all spoke English and were quite familiar with our earth names for these places. They also had impressive, sci-fi sounding names for themselves, such as "Valiant Thor."[6]

The different species of aliens, which range from the tall, attractive, and good-soldier "Nordic" type which espouses clean living, exercise, and spirituality – this is the species which Hitler is claimed to have believed spawned the Aryan race[7] – to the small, spindly, and physically repellent gray type which is believed to be at best amoral, but is in any event entirely devoid of emotions, to huge, lizard-like reptilians and various other monsters.

The important thing to note here is that each alien species is not only strikingly distinctive, but represents a familiar creature "type," and that these "types" correspond exactly with what we humans regard as "archetypes."

In other words, they seem to be exactly what we would expect aliens to be. At any rate, the assorted alien

---

[6]   See Appendix C for a summary of the highly contentious claim that an extra-terrestrial from Venus named "Valiant Thor" arrived in America during the Eisenhower administration and lived in a secret apartment at the Pentagon, as well as a summary of some of the conspiracy killings associated with disclosures about information regarding extra-terrestrials.

[7]   See Appendix D for a summary of Hitler's obsession with UFOs, and the extra-terrestrial connection to his racist agenda.

archetypes which assembled here on earth immediately after the first nuclear test also claimed to be warring with each other – also exactly as we would expect of aliens species from other planets. They then threatened to take over *earth* – once again exactly as science fiction has conditioned us to expect.

But because they were at odds with each other, the aliens offered *our* "leader" competing deals, the most generous of which, predictably, came from the benevolent, god-like Nordic species, who volunteered to defeat the bad-guy Grey species and protect earth from general alien invasion – if *we* would agree to nuclear disarmament.

This deal was rejected, as they must have known it would be, given even a cursory observation of our planet. But perhaps it was worth a try, and represented a sort of "Plan A."

The deal which is alleged to have been accepted involved the American military receiving ET technology from the "Grey" species in exchange for visiting gray aliens being permitted to remain on earth, hidden from view, and engage in monitoring, via high-tech implants, and tissue extraction and experimentation on humans – who would be abducted entirely against their will, but returned unharmed, with no memory of the abduction.

Exactly *how* these aliens were supposed to "erase" human memories of their abductions is not clear, but the Greys were also given permission to experiment on animals – specifically, on cattle – from which they would surgically extract tissue and organs for unspecified purposes.

The above synopsis relating the origin of the current alien presence on earth is admittedly complicated, but the main thing to note here is that either we humans derived all of our fictional archetypes about gods and angels and monsters from observations of visiting aliens which looked like these things, as it is argued in the UFO literature, or – as will be argued here – that when aliens come to visit earth, they deliberately fashion their corporeal forms after the sorts of human-created fictional archetypes that they have observed us create, in order to appear "as expected."

Given that these aliens all arrived at the same time, already adapted to our climate and able to speak our language, and that their "types" correspond almost exactly with the archetypes one encounters in earthy images throughout the ages, it seems most probable that they did not come from disparate planets too far away for us to be able to confirm or dispute, and that they are not "warring with each other," but in fact represent a combined effort to get us to disarm – which evokes every single earthly stereotype going back all the way to the beginning of recorded human history.

In other words, these visiting aliens look and sound exactly like our preconceptions about aliens because they have fashioned their costumes after our customary fantasies about them. They "communicate" with us by parroting our beliefs back at us. And they can survive comfortably in our atmosphere because they are not "alive" in the usual sense of that term.

This is not to insist that it is *impossible* that at least some of the purported extra-terrestrial visitors to our planet might be real flesh and alien blood beings, who originated elsewhere in our own dimension.

But it *is* impossible to explain how such beings could even get here, let alone arrive on earth so well-adapted to our climate and culture.

And even harder to understand why they would take the time and trouble to come all the way here simply to bring messages of peace and love to what are essentially a bunch of incorrigibly greedy and angry apes intent upon blowing each other up – sapiens who will not, furthermore, even consent to believe in the *existence* of aliens, let alone listen to what they have to say.

What else could these aliens possibly be but robots? They don't eat, they don't sleep,[8] they don't reproduce, and they don't – apart from the ones recovered from ostentatiously-staged "crashes" – die. They don't even age.

According to some reports, the "Grey" aliens are beginning to lose a bit of their luster – exactly as one would expect of machines – and sort of "disintegrate," as they succumb to the forces of entropy over time. But they certainly don't die in the usual sense; they can't even be killed, as numerous back-country reporters with shotguns can attest. They also don't have emotions, or seem to mind being relegated to underground bunkers for decades.

---

[8] Apparently, aliens claim that the need for sleep was eliminated ages ago, and instead of wasting time "mimicking death," they simply "regenerate" – in some incomprehensibly high-tech manner. To the believer, the ability to go without sleep demonstrates the evolutionary superiority of extra-terrestrials. It also tends to support the contention that they aren't really alive – at least not in any meaningful sense.

And from where could these aliens possibly have come, if not from a near dimension which has been observing and compiling information about us? For space explorers, they seem to exhibit a peculiar lack of interest in anyone or anything about our planet – except, of course, cows. And for visitors from extremely distant places, they sure knew an awful lot about us before they even arrived, including how to dismantle our technology, and how to speak all of our languages. The same aliens have been here for over 70 years now, hiding underground and ... Doing *what,* exactly? What could highly-evolved, sentient beings possibly be doing in hidden sub-basement bunkers for all of that time to amuse themselves?

If these alleged inter-galactic aliens are intent upon colonizing, why haven't they taken at least a couple of preliminary steps in that direction by now? And if they are in fact controlling our government, as most ufologists claim, and their primary purpose is to save and/or protect our planet from destruction, why haven't they managed to do a better job, or at least introduced some really robust environmental legislation? And what about the religious component? Is it even remotely possible that at least some of these ETs really *are* inter-galactic missionaries?

If they truly are sentient space explorers of some sort, this would seem to be the only feasible explanation. Religion, recall, is the one belief system which is common to every civilization, in every era. Why should we automatically assume that aliens are exempt? And consider further: Who is the individual most likely to volunteer to leave the comfort and familiarity of his home, braving all manner of boredom and danger for the sake of a mission? The missionary, of course. But whether or not they are religious for real, it would be absurd to declare that these aliens do not *exist*, simply because they espouse religion.

When, in previous centuries, earthly missionaries suddenly materialized in hitherto unvisited South and Central American countries, determined to bring their message to the natives, they were immediately – and, contrary to contemporary practice, reasonably – identified as "aliens."

Some indigenous cultures further identified them as gods, and some went so far as to identify them as the gods they had been waiting for since time immemorial. But none of them, it seems, declared that these pale, alien, missionary people did not *exist,* simply because they didn't buy their message.

Why scientists would do such a thing today is a problem more for sociology than for logic, but it does seem clear that if extra-terrestrial aliens have arrived on earth claiming to be bringing their religion to us, the reasonable thing to conclude is that extra-terrestrial aliens have arrived on earth, claiming to be bringing their religion to us. Which brings us to the next salient point.

Why would alien visitors with a message insist upon hiding themselves, land in isolated areas where hardly anyone would be expected to interact with them, and attempt to make deals with government officials by threatening destruction unless their presence remained concealed? What can an entity capable of paralyzing an entire geographic region by knocking out its power grid telepathically possibly have to fear?

My suggestion is that an entity which is capable of knocking out a single power grid telepathically might be quite helpless when it comes to simultaneously controlling the multiple individuals contained in crowds, and might be extremely vulnerable if captured. Moreover, if its auxiliary agenda involved obtaining terrestrial genetic material – including that of cows and humans – turning out our lights temporarily is not going to accomplish this end, however much it might succeed in impressing and/or frightening us.

It is alleged that aliens have extraordinary powers to telepathically control individual humans, immobilizing them and even transporting them telepathically upwards and across great distances to craft where examinations are performed, and tissue and bodily fluids extracted. Presumably, they haven't been doing this for 70 years simply out of curiosity.

If aliens have come all this way in order to perform such procedures, then their need must be dire. And indeed, the representatives of the species of aliens involved in this practice purportedly claim that they are dying due to over-cloning – their only available means of reproduction – and so desperate to replenish their genetic stockpile that they are even willing to use human sources.

This claim is believed, repeated, and upheld in all of the UFO literature, where it elicits a great deal of sympathy for the poor dying race, and its homely little gray representatives who brave the perils of inter-galactic travel – often only to die themselves in tragic crash-landings in our terrible deserts. But is it really reasonable to believe that such aliens are capable of assuming control of our entire planet, even if they *can* immobilize and abduct individual humans telepathically? How, exactly, might this be possible?

Putting aside for the moment what precisely it is, and indeed whether it even *exists*, it seems reasonable to assume that telepathic "strength" would be akin to physical strength, and limited in more or less the same way. An individual's prowess when it comes to performing amazing feats of weight-lifting, for example, really only works on a one, or two-weight basis, and for a limited period of time. A bench press, in other words, involves lifting a single set of weights and then putting it back down – not hoisting up every barbell in the land and holding them all up in perpetuity.

So perhaps an ET which can easily control a single human individual telepathically, for a limited period of time, would be helpless in the face of an entire crowd, or even a small group. They might be smarter and more technologically advanced, but we still outnumber ETs, and it would almost certainly be impossible for a few of them to abduct or otherwise control us all at once – forever.

Nevertheless, impressive temporary displays, combined with the threat of destroying our energy sources altogether if we refused to comply with their demands, might be sufficient to induce a government to make a deal – as whistle blowers claim that President Eisenhower did[9]

---

[9]   For a summary of Eisenhower's alleged interaction with extra-terrestrials, including the signing of an inter-galactic treaty, see Appendix E.

– allowing the gray aliens to mutilate a controlled number of cattle and abduct human subjects for tissue extraction and surveillance implantation on a limited, strictly-defined basis which guaranteed that such individuals would be returned, unharmed, and with memories wiped clean.[10]

But even if the information concerning the Eisenhower treaty turns out to be nothing more than an elaborate hoax, perpetrated by hundreds of seemingly unrelated persons for reasons unknown, the number of cattle mutilations has, to date, escalated almost into the tens of thousands, and many abducted humans have simply disappeared, or have been returned with memories fully intact, terrified, and psychologically destroyed for life.[11]

---

[10]   The so-called "Greada Treaty," purportedly signed in 1954.

[11]   The number of missing persons in America at any given time is estimated to be 90,000. Many of these people simply "vanish without a trace," and in the UFO literature such disappearances are often attributed to ET abductions.

More specifically, the number of children who "vanish" while their families are camping in isolated nature areas or visiting national parks is alarming: They haven't been kidnapped, their bodies are never found, and nobody seems to be able to account for any of them.

On extremely rare occasions, a disappeared child will suddenly re-appear days later, in an inaccessible, far-away location, sometimes in sub-zero temperatures, perfectly healthy, without a scratch, and with no memory of where they have been. More usually, an article of clothing subsequently appears in a conspicuous spot which has been already been searched, but no trace of either the child, or any trauma, is ever found.

Whether or not this fact is related, many abductees claim to have seen/interacted with groups of hidden children, and while they do not deny the truth of these claims, the pro-UFO camp

But despite all of this, the reflexive pro-UFO reaction espoused in the literature is to rationalize *all* ET behavior, and attempt to avoid being anthropocentric, or assign "blame" – especially when the transgressor is a vulnerable, lost-looking little alien whose species is in peril.

Yes, the "Greys" violated the Eisenhower treaty, ufologists concede; indeed, they violated it immediately, and egregiously, too. But it would be a mistake, the UFO sympathizer sympathetically reasons, to assume that little lost aliens subscribe to the same human-specific morality that is embedded in the sort of deal-making which formed the basis of Eisenhower's treaty. They come from a completely different world, after all, and can't possibly have understood what was meant by "keeping their word!"

Yes, they have violated humans – bodily, indecently, and repeatedly, the sympathizer concedes. But extra-terrestrials come from naturally "communal" civilizations, and are essentially incapable of even comprehending the concept of individual autonomy being infringed upon when they render their abductees immobile, insert high-tech spy

---

insists that the military is behind "most" of the child abductions – experimenting upon and even breeding humans to produce a race of "super soldiers" who are mind-controlled to defeat the "ET menace." Anonymous super soldiers have apparently confirmed their existence, and even claim to have travelled across dimensions to battle hostile alien monsters.

It is all so fantastic that one does not know who or what to believe, and the issue is complicated further by accusations that the military is deliberately leaking false "disclosures" in order to protect their UFO and space weapons secrets by making ufologists sound ridiculous. Whoever is disseminating this information is certainly succeeding.

bugs, and extract bodily fluids and fetuses from their earthly friends!

Moreover, an extra-terrestrial species which communicates telepathically, the UFO sympathizer reasons, would lack the concept of privacy, as its very *thoughts* are automatically public. And its ability to perpetuate its race by cloning would make assorted human taboos surrounding reproduction, and the human abhorrence of human hybrids as "grotesque," similarly incomprehensible:[12] But success in accomplishing a human/animal hybrid is sufficiently exciting, surely, to excuse any of these violations!

Finally, the UFO sympathizer points out that such a highly-evolved alien species would naturally view humans as insufficiently conscious to warrant moral consideration even if it did possess this uniquely human concept, and were not acting under the compulsion to save its own race by borrowing a bit from ours. Besides, it won't kill us to share!

All of the above justifications, are, to be sure, entirely well-taken, given a fellow creature's great peril. But granted that he does truly exist, and does truly make claims about being in danger of extinction, are any of the extra-terrestrial's claims really credible? Why, for example, don't these allegedly-endangered little aliens simply make use of animals on their own planet, if they

---

[12] One ET booster even rationalized the creation of ancient human/animal hybrids as something which was done for human benefit, describing them as "awesome." This is the correct usage of that term.

need to create ET/animal hybrids for the survival of their species? Are they the only species living on their planet? If so, how very odd. And what, for that matter, exactly is "over-cloning," and why would such a procedure cause problems for this singular species? One would assume that the process would become increasingly perfected over a long period of time, rather than the reverse.

And finally, how is it even possible for a species to perpetuate itself entirely by cloning? Where did they come from in the first place? Did they suddenly emerge out of cabbage leaves on their strange, single-species planet, fully-formed, hyper-intelligent, and conveniently equipped with laboratories designed expressly for this purpose?

But even ignoring all of the above improbabilities, *and* accepting the idea that they could, as they claim, somehow "inject their own consciousness" into a (horror alert!) human/cow hybrid – which is essentially preposterous in any dimension, given the nature of consciousness – would such a highly-evolved, hairless, nattily-attired aluminum little type, so opposite in every conceivable way, really consent to perpetuating its species via a furry bloated abomination like a human/cow hybrid, even if they *were* somehow able to transport these huge, malodorous monsters to another star system, or galaxy, or wherever it is that they are supposed to be coming from?

Yet this is the story which is unquestioningly believed and repeated without a hint of horror – or humor – by ufologists: That tiny, put-upon, gray aliens, exactly as portrayed in the film "ET," *have* to abduct and extract human tissue and mutilate cattle in order to create human/cow hybrids in which to inject their own consciousness so that they can save their dying race on a distant planet!

It sounds so outrageous that one's natural inclination is to simply deny that such aliens exist. But again, it would be well to keep in mind that the truth of ET's *claims* is a quite a separate matter from the truth of his *existence*, and it is quite possible to dismiss the former while still acknowledging the latter.

Since the evidence for the existence of extra-terrestrials would seem already to be almost overwhelming, and it increases daily as the government releases a slow, steady, stream of de-classified images[13] and documents and the internet goes into overdrive with amateur videos of everything from underwater alien sea bases to adorably retro little flying saucers, the most reasonable thing to conclude is that these aliens exist; they abduct humans; they mutilate cattle – and they are not especially honest about who they are, where they are coming from, or why they are here.

As to their exceedingly questionable claim that they need to extract and experiment upon earthling tissue in order to alter (and thereby "save") themselves, the most reasonable explanation for an alien species wanting to tinker with human genetic matter in this – or any other – way, it would seem, is that their intention is to genetically alter *us*. And, taken together with their insistence that we dismantle our nuclear capability, the most reasonable

---

[13]   To view samples of government-approved UFO "disclosures" from the 1940's to present, see Appendix F.

explanation for ET visitors wanting to genetically alter us is that they desire to prevent earthlings from annihilating their own dimension along with ours, but for all of their technical sophistication, they do not possess the means to control us directly.

Consider: These "desperate" aliens arrived with the first atomic blasts, and, despite their pleas for help in saving their dying planet/species, seem primarily interested in our nuclear technology. They threaten us with the ability to destroy our electrical grid, and demonstrate their "powers" by turning out lights in limited geographic areas, or briefly disabling nuclear-tipped missiles "telepathically."

Wherever they are coming from, if nuclear explosions here have injured their habitat so badly that they have arrived on earth, *en masse*, to prevent us from ever using this technology again, and they in fact possess the ability to disable our global nuclear arsenal permanently, why don't they just do it? We want the same thing and are clearly incapable of accomplishing the task ourselves; why don't they help us?

If the situation is that critical for them as well, and they truly possess advanced powers, ETs would not simply *threaten* to disarm us, or attempt to coerce us by issuing gentle persuasive sermons about the serene and peaceful oneness of all existence and their commitment to the imperative of human autonomy and the leafy enlightened lifestyle. They would dismantle our weapons – *all* of them – and assume control of our governments, our armies, and our education systems.

If, on the other hand, they did *not* possess the ability to solve the problem directly, and they had to somehow convince us to disarm ourselves, what they would do is inundate us with a bewildering variety of ET forms, and try everything: Try tempting us with nirvana; try altering our DNA by hybridizing it with other animals; try terrifying us with the results; try bribing us with offers of advanced ET technology, and, finally; try preying on our sympathy by crying for help with their poor dying species back home, which cannot reproduce except by cloning – us. All of these techniques have proved effective, incidentally; the only real failure here would seem to involve veracity.

It seems much more logical to assume that the real reason that this ET species is not able to reproduce biologically is that it is neither a species, nor – or at least not wholly – biological.

It also seems more logical to assume that all of the assorted "crashes" and "deaths" of these tiny, vulnerable aliens are simply staged, in order to evoke our sympathy and gain access to our military facilities, where they will subsequently be able to gather information via embedded micro-technology. And finally, it seems likely that the development of hybrids is itself little more than a sickening stunt to bend us to their will – even if it is only a ruse.   There is, unfortunately, evidence to suggest that it is not.

Apparently, such hybrids not only exist, but have been openly displayed to select human abductees for the specific purpose of causing them to report their experience – the hope being, presumably, that the general earth population will not only believe the reports, but believe that these aliens have the capability to transform our entire species in such a way.

The existence of alien/human hybrids is generally not in dispute. Virtually everyone in the ufology community also takes it as given that aliens have the ability to radically alter and/or control the entire human species in this way. But if this is the case, why don't they simply do so, and transmute the lot of us into contented, cud-chomping human cows, with the bovine standard of non-aggression? Why go to all the trouble to secretly create human/animal hybrids which, however much they might terrify and repel us, can do little more than die,[14] since hybrids are sterile? Or, more efficiently, why not simply kill us all outright? ETs exhibit little compunction when it comes to killing individual humans who interfere with their agenda or

---

[14] Another possibility is that the hybrids are being created for political purposes. In sufficient numbers, they might be able to organize politically if, as it seems we would, Americans would want to give them equal rights. For its part, the Vatican has already expressed this desire; that is, it has expressed a willingness to accept *all* extra-terrestrials into the Church, and presumably this would apply to hybrids as well. At a Vatican conference regarding this issue, the Pope's chief astronomer, Father Gabriel Funes, announced:

> "Just like there is an abundance of creatures on earth, there could also be other beings, even intelligent ones, that were created by God. That doesn't contradict our faith, because we cannot put boundaries to God's creative freedom. As saint Francis would say, when we consider the earthly creatures to be our "brothers and sisters", why couldn't we also talk about a "extraterrestrial brother"? He would still be part of creation."

attempt to expose them; why not simply eradicate us as a species? Isn't that how evolution is supposed to work, after all? If their claims to ultimate authority over our planet's evolution are true, they've done this sort of thing before.

And if, as ufologists insist, there really *are* diverse alien bio-species living in underground earth cities, who have been here all along, and have advanced powers, why haven't they emerged long before now and used those powers to help, or at least to control us? Why are they, like the purported space-based aliens, so afraid of us?

It might, as these aliens allege, be that all of them belong to the same universal benevolent society which consists entirely of species which always behave collectively, for the greater good, but are nevertheless so committed to the concept of *human* autonomy that they have hitherto refused to even comment about our paths of destruction – despite the fact that we are destroying their habitat as well, and so unbearable as neighbors that they have gone underground to get away from us.

But it seems much more likely that these beings, wherever they are coming from, are all coming from the same place, and coming only under extreme duress. And that, far from being able to control unruly humans, they, like us, are essentially powerless in the face of nature, and are still learning about our respective spots in the universe – despite perhaps having been *aware* of us all along.

They might, as some anthropologists contend, have even visited our earth in earlier eras. But instead of visiting for the purpose of *gifting* humans with advanced technology, as such anthropologists further contend, given that their arrival seems invariably to have corresponded with the presence of ancient advanced technology, it is much more likely that ancient ETs were *obliged* to visit our earth – in times when terrestrial civilizations had, due to the proliferation of advanced technology, also reached the point where the military application of this technological expertise threatened to cause massive, dimension-spanning levels of destruction.

They might even have attempted to contain the problems posed by these ancient technology-loving humans in exactly the same way that they are attempting to contain the problems posed by us: By promises of spiritual rewards for compliance, bio-terrorism in the form of repugnant human/animal hybrids, and threats of total annihilation. Just how literally we are justified in taking classical accounts and ancient artwork depicting such events depends upon one's point of view. But certainly the current near-repetition of historically-recorded classes of similar occurrences – however unscientifically documented – would seem to add, rather than to detract from the probability of their truth.

It would also seem to cast serious doubt upon the contention that these visiting aliens *provided* ancient civilizations with their technology, but this is a separate problem.[15] The point to note here is that at some time in

---

[15]   It is interesting to note that what often appear to be "impossible" leaps in scientific knowledge and/or technology and are therefore attributed to alien sources really only seem that way if one subscribes to the theory of evolutionary "gradualism" which is embedded in our current world-view.

our own recent past – generally agreed to be when we detonated our first atomic bomb – higher-order entities became aware that human technology was a menace to the greater universe which we all inhabit, and either arrived, or increased their presence here in a manner which is both alarming, and escalating.

And, as with human exploration of the micro-worlds contained within our own when we discovered how to split the atom, their knowledge of our shared universe increased with increased observation, and enabled them to construct tools which permitted them to not only observe, but to alter aspects of our habitat, albeit in an analogously limited sort of way. But such ability creates, at best, the *illusion* of control – even if ETs have, as is more probably the case, been observing us all along, and have simply not found us sufficiently compelling to study in any great depth.

———————

Knowledge, like evolution, is not a gradual procession towards the ultimate expression. It advances in leaps and bounds, and sometimes a single discovery is all that it takes to launch an entire techno-revolution: Witness the intellectual explosion which followed the discovery that the atom was divisible, or its theoretical precursor, the identification of heat as a form of kinetic energy – providing as it did a new foundation for all contemporary physics with its paradigm-shattering expansion of the concept of "energy."

We might be undergoing a similar revolution in scientific understanding now. Or, as ufologists insist, it might be that someone is getting their information from aliens. But proof of this claim requires more than a mere expression of awe in the face of technological advancements, and a "Where else could *this* possibly come from?"

There were, it is generally accepted, technologically advanced societies of humans on the planet long before ours. There have also, according to scholars in a variety of classical disciplines, been reports of visits from beyond, freakish human/animal hybrids, and otherworldly threats of near-total annihilation – some of which were carried out. The ufologist community has, for the most part, embraced recent (highly controversial) literal interpretations of these scholarly reports, and the existence of contemporary aliens of similar appearances certainly lends credence to this position. Given the remarkable parallels between these ancient tales and the present-day ET "invasion," it seems at least possible that these long-lost ancient civilizations faced alien hostility essentially equivalent to what we are presently experiencing when they also commenced, as seems to be human nature, carrying their advanced-technology destruction to the point where it threatened to undermine the stability and comfort of the greater universe.

The further contention that these ancient advanced-technology civilizations *received* their advanced technology from ETs, on the other hand, would seem to be extremely improbable, given even the briefest observation of the former, by the latter.[16]

---

[16]   One account claims that Sumerian knowledge of a specific celestial event which they could not have observed because it occurred before they came into existence "proves that they got their information from ETs."

To set the argument up logically, it would look something like this:

Yet it does seem reasonable to assume that the ancient aliens had been observing ancient humans all along – out of curiosity, initially, perhaps, and finally out of horror and apprehension – just as they have apparently been observing us, and probably only sent their emissaries down here *en masse* when drastic action was required to protect their own interests, exactly as is the case today.

Ancient ETs might well have originated in a higher dimension, as is often claimed by ufologists, and, this being the case, their relationship to humans might be analogous to our relationship with the micro-processes in our own bodies.

A handful of scholarly individuals have always found these internal processes interesting in themselves, and dedicated themselves to studying them for the sake of science. But for the most part, humans find the entire business of their internal engineering dreadfully boring and/or distasteful, and are quite content to both remain ignorant of its mechanics, and ignore all manner of nagging little

---

1. "Sumerian society had not been in existence long enough to observe celestial event (X).

2. Sumerians had knowledge of celestial event (X).

3. Therefore, Sumerians acquired their knowledge of celestial (X) from some source other than direct observation.

Googling "Pre-Sumerian civilizations" provides a couple of distinct possibilities. But note that this is not necessarily a matter of acquiring the information from some other human source; it is also possible to predict/derive information abstractly, from purely theoretical considerations. This is not to suggest that the Mesopotamian ET thesis is impossible, but only that it is not probable, let alone "proved."

symptoms that something is amiss in there for as long as they possibly can before seeking a solution.

When the situation becomes critical, and our deteriorating health interferes with our ability to go about our normal business, we are forced to address the problem and take – often drastic – action to contain or control it. But even in cases where it is possible to do this in a limited sort of way, absolute control over our internal processes is inherently impossible, regardless of how advanced our medical technology might be. Our bodies are hard-wired to run down over time, and ultimately, there is little a doctor can do beyond observing and reporting back with the prognosis.

Presumably, the higher-order entities whom our nuclear destruction has affected are in exactly the same situation: Their corner of the universe – which unfortunately contains us – is likewise running down, as per the universal law of entropy. To keep it going in the best order possible, and for as long as possible, they do regular check-ups to observe its various potentially-troublesome internal components (Read: "Humans").

And although they might have been monitoring us in this way all along (via bases observed on the sea bottom, for example, or on the moon) and possess technology capable of temporarily suppressing isolated pockets of internal-component destruction – that is, controlling individual humans for limited periods of time, in a limited sort of way – if we insist upon testing and using weapons so destructive that their carnage cuts across dimensions, we are going to kill aliens too, and ultimately there is little that the higher-dimension entities affected can do but observe.

So they manufacture surveillance tools and earth-appropriate laparoscopic probes and "inter-planetary visitors" in the form of human language-mimicking bio-robots in the desperate attempt to find out what is so amiss with us down here, that it is affecting *them,* up there.

Their ability to perfect these instruments, like any applied science, increases – often dramatically – with the increased understanding which observation brings, and the bio-robots become increasingly realistic-looking, and increasingly able to communicate effectively with us.

Our present build-up and testing of nuclear devices has, at this point, become so critical to the general health of the greater universe in which we are contained that ETs have apparently posted observational devices everywhere in our atmosphere, and even, if qualified observers are to be believed, on neighboring planets.

Astronauts, it is reported, have seen the very same sort of UFOs on the moon that millions of people have reported seeing on earth, and these reports are increasing daily. But it doesn't follow from this – or at any rate it does not follow *necessarily* – that there are advanced civilizations on the moon, and that they have been there all along.

Bio-inhabitation of a planet involves the existence of diverse life forms and the habitat to support them; in its absence, the presence of UFOs, even in abundance, suggests nothing more than the presence of UFOs – in abundance. Nevertheless, they *are* there. So why do scientists deny it?

Historically, in a healthy scientific paradigm the appearance of new phenomena is explained by extending already-established general laws. The appearance of UFOs represents new phenomena, but they cannot be explained by extending any scientific principles available in the currently-accepted body of scientific knowledge. This is why scientists deny their existence: Given presently-accepted general laws, the things should not be here. But, incredibly, instead of concluding that therefore their scientific laws must be inadequate, scientists conclude that UFOs do not exist.[17] They then proceed to make jokes about them.

The "new-paradigm" UFO adherent, equally incredibly, concludes not that current scientific laws are inadequate to explain the presence of UFOs, but that the presence of UFOs discredits the legitimacy of science *itself.*

---

[17] J. Allen Hynek, an astronomer turned ufologist, is also a wonderful writer whose work is peppered with entertaining examples of this and other logically fallacious inferences made by scientists. The one attempting to explain away a flying saucer observed by hunters by hypothesizing the existence of an independent platter thrower driving to the woods for no other reason and hiding behind a tree in an otherwise deserted forest where hunters arrive at precisely that spot at the same moment and, upon witnessing the plate toss, claim to have seen a UFO, is not to be missed. And it gets even better when the platter comes back.

He also relates an anecdote to illustrate the "Whatever I can't explain does not exist" fallacy in the form of a poorly educated zoo patron who, upon observing an exotic species with which he is not familiar, exits shaking his head and muttering, "There ain't no such animal!"

He then equates science with materialism, and, the strictures of both having been gotten out of the way, finds himself free to speculate about galactic-spanning spirituality and the possible connection between extra-terrestrials and the various awe-inspiring creatures depicted in mythology.

Liberated from the confines of materialism – he uses the terms "scientific" and "scientific/materialist world view" interchangeably[18] – he claims the right to pursue his interest in plasmas, ghosts, poltergeists, and the possibility of things like disembodied consciousness, shape-shifting, and re-incarnation. All of these things, it is asserted, are associated with ETs, and *they* exist, so why not?

But even if one were to grant the truth of any or all of the above, the problem of how ETs from distant – or even nearby – planets are able to arrive here both biologically compatible with our atmosphere, and able to communicate with earthlings, remains. And *its* possibility has a probability approaching zero, even apart from the difficulties involved in explaining how the UFOs which they allegedly pilot could get here in the first place, given the tremendous distances involved. To tackle this problem, ufologists tend to take one of two, diametrically opposed, approaches:

------

[18] The identification of science with "materialism" is a common misperception outside of the UFO literature as well. But there is nothing about science which makes the metaphysical theory of materialism necessary. Ontological materialism is simply another theory – and, like all theories, it can be disputed and/or replaced.

1) They might, as many hard-core science-fiction-minded adherents do, steadfastly maintain that UFOs are genuine, "nuts and bolts" machines in the good old-fashioned sense, and imagine high-tech, anti-gravity propulsion systems to get them here, with electro-magnetic shields, possibly, to protect them from adverse effects as they zoom through outer space at three times the speed of light.

Teleportation, too, is a possibility, as is the existence of alternate, parallel dimensions as their places of origin – permitting UFOs to do things like bend space/time and slip through vortexes and tunnels and Stargate portals to arrive here instantaneously, and possibly from the future.

This ignores the improbabilities associated with ET bio-compatibility once they arrive, but it does offer a variety of possibilities for getting them here, all of which, in addition to being extremely exciting, are apparently permitted by contemporary physics, and even, allegedly, being secretly tested in covert military operations. Or they can;

2) Postulate alternate, parallel dimensions as places of origin, but imbue them with occult, science fiction-defying properties like body-less plasmas which can levitate objects and walk through walls, universal souls, communion with ghosts, spirits, and angels, and collective consciousnesses.

Both camps further postulate a universe filled with populated planets – underground, in planets close enough to observe from here – containing all manner of alien species arranged into hierarchical governments, alliances, and various warring factions.

The occult camp sometimes envisages our earth as a sort of hell to which transgressing souls are banished as punishment until they can be re-incarnated into one of the higher realms, and the sci-fi camp sometimes describes it as a portal to other worlds through which all manner of aliens are required to pass in order to access higher dimensions. Both tend to assume the necessity of openly integrating with the aliens already here, and some assume that a large influx, or "return" of at least one dominant ET species is imminent.[19]

Because all of this sounds so fantastic to someone trained in epistemology – especially to someone who, like the author, has only recently been initiated – and is entirely too much to assimilate, let alone critique, in what is left of a limited lifetime, critical analysis of ufology will be confined here to the claim that there are intelligent alien species on our earth who have arrived here from other planets and/or parallel dimensions.

To begin, consider the oft-advanced claim that our universe contains "parallel dimensions." Is there anything

---

[19]   The species proposed is generally the "Nordics" – the *Viking-esque* race with which Adolf Hitler identified. According to some sources, these alien Nordics now inhabit regions of northern Scandinavia, near the North Pole; according to others, aliens were crucial in providing Hitler with the advanced technology which they believed would permit him to win the war, and the German scientists who emigrated to the United States after it had been lost continued to engage in alien/Nazi-inspired experimentation into occult subjects like mind control, telepathy, and eugenics.

in the currently-accepted scientific paradigm which would render the existence of parallel dimensions to be expected, or are they simply exciting mathematical constructions with no corresponding empirical import?

If the latter, then they do tend to give the appearance of being "empirically irrelevant," and the proliferation of purely abstract, impossible-to-confirm hypotheses like this is another symptom of a scientific paradigm which is past its prime. [20] But as a proposal for a *new* paradigm, as so

---

[20] "Theories of everything," and "strings in space" are typically advanced – and received – as paradigm-shattering discoveries, but to the extent that they are accepted, they would appear to fall squarely within the current scientific world view. The very idea of reconciling quantum and cosmic theory suggests a continuation of, rather than displacement of the present paradigm, and adding another dimension or ten to the cosmos does not alter our understanding of it appreciably, especially if the dimensions in question have no empirical significance, and can scarcely be conceptualized. Mathematical constructs of this kind are exactly the sort of thing that currently-accepted science congratulates itself upon, but the epistemological underpinnings of these theories are positively reactionary in nature.

Consider, for example, the historically important, but now obsolete philosophical concepts of reductionism and unitary science. Both were abandoned as unfeasible generations ago, and a unified field theory – even it does succeed in reconciling quantum mechanics with relativity by formally "re-defining" the fundamental structure of matter – will not solve the underlying philosophical problems which would seem to make the entire program impossible even in principle.

This is hardly the fault of scientists, who cannot be expected to have a background in philosophy. But the above focus overlooks another, even more fundamental, discrepancy in science today, and that is the inherent difference between biological and

much ufology purports to be, the synthesis of New Age concepts like channelling and the (often intentional) mirroring of popular science fiction, combined with an unquestioning acceptance of anything communicated telepathically by ETs through psychics and "mediums," and then filtered through the presuppositions of American Judaeo-Christian ontology and morality, makes for an uneasy alliance at best.

Of course it is *possible* that earth's diverse alien visitors are real biological entities who are truly able to communicate with humans in some advanced, "telepathic," way; have infiltrated and influenced our government in order to halt our proliferation of destructive nuclear technologies, and; are telling the truth when they claim that the cosmos really *is* arranged into an hierarchical, Star Trek-type order devoted to universal peace and enlightenment.

It is *possible* that every single planet and/or parallel dimension in the universe consists of species immeasurably more advanced than earthlings, and that they all have the power to control humans absolutely, but simply choose not to exercise their powers because they also possess Judaeo-Christian notions about freedom and

---

physical laws of nature. It must be questioned why any project would proceed on the assumption that the former can somehow be "reduced to" the latter simply (well, metaphorically speaking) by means of positing a more complex and/or expansive quantum theory, when it is generally accepted that, given not only the logical differences between the two, but the functional – or essentially "teleological" – nature of biological processes, this is impossible even in principle.

are dedicated to the principle of letting us learn our own lessons. But it does seem unlikely, especially since the one characteristic which seems to be common to all of the alien "species" reported as visiting/inhabiting our planet, is their fear of being discovered.

In almost every report, the need for secrecy is ET's paramount consideration, and they will harm or kill individual humans as necessary, simply in order to avoid detection and exposure. They are clearly concealing something, and the most probable culprits are:

1. Their alleged ontological status as flesh-and-blood and, if the new paradigm is to be accepted, immortal-soul-bearing, sentient beings, and;

2. Their alleged power to control and direct humanity.

Because their power and influence over humans is said to extend to our very evolution, this might be a good place to begin. The various attempts to attribute human "creation" to the ancient hybridization of advanced aliens and lower, earth-based animals (i.e., Bigfoot) is, while certainly possible, extremely unlikely as the sole source of our evolution. Granted that (if depictions from ancient Sumerian, Egyptian, and American Indian cultures are to be taken literally) ET and/or human-animal hybrids have existed in the past.

But presumably, like all hybrids, they were sterile, and did not even perpetuate *themselves*, let alone entirely new, dominant species. Why would they suddenly be able to do so in the case of ET/human hybrids – and *only* in the case of ET/human hybrids? And how does this explain how all of the other bio-forms arrived on earth? Did they evolve in the old, natural selection way, and then suddenly morph into man via injection from ETs in the relatively recent past?

According to ET-based evolutionary theory, which simply ignores all other bio-evolution, and indeed cosmic evolution in its entirety, there is no other possible explanation for human intelligence. Sumer civilization, it insists, which lasted for a scant 2,000 years, simply could not have produced its spectacularly-advanced scientific predictions and technology without help from ETs.

And perhaps ETs really *were* the source of the remarkable knowledge displayed by ancient advanced civilizations, including Sumerian technology – the blueprints for which are believed by some ufologists to be hidden in what is now southern Iraq, and are even alleged to be the real source of wars and conflict in that area, as covert, international military operations conspire and compete to unearth the buried technology in order to use it for advanced, ET-inspired weapons production.

But it seems much more likely that it is the highly questionable, "2,000 year" Neo-Darwinian geological time frame presupposed here which makes ET knowledge-injection seem like the "only possible" explanation for the advanced technology attributed to ancient advanced civilizations.

There is an abundance of credible empirical evidence to demonstrate conclusively that man in his present human form is not, as Neo-Darwinians claim, a recent arrival on the planet at all, but has been here sufficiently long for these "analogous" ancient civilizations to acquire their knowledge in exactly the same, cumulative sort of way that we did. The real puzzle, it seems, is not how these early humans acquired their knowledge, but why the civilizations possessing it all mysteriously collapsed.

This is a subject of considerable current speculation and scholarship, and the question of how bio-evolution itself was accomplished remains open, given the difficulties associated with Neo-Darwinism. But it is clear that the ET hypothesis – at least on its own – cannot solve this puzzle, because ET influence is supposed to be limited entirely to human evolution.

In other words, even if humans did *not* evolve from apes via a slow, gradual process of random genetic variations plus natural selection, but were "created" by alien hybridization with terrestrial animals, where did the animals come from? But the biggest problem with the ET hypothesis in general, it seems, is what might be referred to as the "irritation factor," and it infects both camps.

Both professional and amateur scientists have a tendency to simply tune out when confronted with concepts like embracing the inner spirit, channelling, energy healing, and the like. Even if these concepts can, at least theoretically, be re-defined in strictly technical terms, they refuse to make the effort – refuse even to *listen,* because the mere sound of them irritates the science-minded individual beyond endurance.

Scientists typically have the same reaction to epistemological analysis, and will cling to a tattered and terribly outdated and even inconsistent set of beliefs – which they will either impatiently *repeat,* when challenged, or simply roll their eyes and make jokes about anyone who "is too stupid to understand" their obsolete, logically-incoherent "scientific" principles.[21] The rationale behind this, of course, is the underlying theoretical presupposition that whatever is called "science," is inherently superior from an intellectual standpoint, so regardless of how bad a particular scientific theory is, it can still be called "science" – and this is precisely what New Paradigm ET proponents object to.

New-paradigm ufologists, for their part, are irritated beyond endurance by the mere mention of what they derisively dismiss as the "scientific/materialist" world view, and are attracted instead to the spiritual, the occult, and the supernatural.

---

[21] The author once attended an academic lecture where two scientists were invited by a philosophy department to present their latest theory, and when the talk was over and they were challenged on the grounds that an unrelated principle which could be derived from their theory was inconsistent with the laws in another, already-established branch of science, the presenters repeated their theory for over an hour, with increasing volume and irritation, while the philosophers restated, rephrased, and finally, shouted what the problem was. At the end, both sides stomped out, with the philosophers exclaiming, "How stupid can scientists be? They can't even make a simple logical inference!", and the scientists muttering, "Boy, are philosophers stupid! Our theory has nothing to do with that branch of science they kept harping about! How many times do we have to repeat ourselves before they get it?"

Identifying UFOs as "supernatural," they flatly refuse to consider the possibility that any explanation of them is possible within the rubric of even an updated science, and not only view ET presence on earth as a confirmation of the superiority of non-scientific reasoning, but espouse what essentially reduces to a worship of extra-terrestrial beings, while rejecting logic, even in principle, as an acceptable mode of explanation.

These two, diametrically opposed camps also view the ET presence on earth in mutually-irritating exclusive ways, with the pro-science camp sometimes devoted to eradicating it by any means necessary, including using its own reverse-engineered technology to construct weapons to repel the alien menace, and the pro-spiritual camp anxious to integrate ET life forms into human society and creating "exopolitical" theories about how best to accomplish this.

individuals who have (often "recovered") memories[22] of having been abducted by extra-terrestrials are similarly of opposing view-points regarding the experience. Although virtually all of them have been traumatized by the experience to the extent that they are not able to function normally, and spend their lives seeking some means of repairing the damage – either through conventional psychiatry, or various brands of religious and New Age healing – not all of them regard the experience as negative.

---

[22] For a brief summary of the assessment/treatment for alien abduction syndrome, which is classified as a form of Post-Traumatic Stress Disorder, see Appendix B.

Many abductees who have discovered strange, high-tech implants, which are presumably surveillance devices of some sort, refuse to have them removed, convinced that they are serving some "higher purpose" – not just for humanity, but for the larger universe. Often such individuals believe that they are "blessed," or have been "chosen," and view their inability to establish normal human relationships or participate in society as an acceptable side-effect.

The anti-abduction camp either dismisses such spiritual interpretations of the experience as deluded narcissistic nonsense, attributes them to brainwashing by extra-terrestrials, or, in cases where devices remain implanted, due entirely to "messages" sent by the implants. To the extent that they address it at all, the broader UFO community, most of whom have witnessed extra-terrestrial events without having been themselves abducted, are likewise divided on this issue.

Some refuse to even consider the possibility that aliens have ever abducted anybody, and either attribute abduction claims to mental illness, or insist that they cannot be verified scientifically, despite the fact that all manner of unidentifiable foreign micro-objects – many of which go undetected by the victim until they are discovered by health care professionals – have been surgically removed from individuals across America, and there are psychiatrists who specialize in this area.[23]

---

[23]   The number of exorcisms has also increased dramatically in recent years, doubling, and even, in some areas, tripling, as disturbed individuals who claim that they are being "controlled," or have been "taken over" seek help to rid

Others speak out adamantly in defense of the practice, as necessary for acquiring information, and argue that the term "abductee" should be replaced with "contactee," due to its negative connotations.

Many, if not *most* members of this camp go so far as to interpret morally repugnant behavior like the abduction of human subjects for the purpose of harvesting and hybridizing human genetic material, as spiritually enlightening for the human victims – conveniently ignoring the fact that these people are sometimes killed in addition to being terrorized beyond repair.

Or, they claim that it is "necessary" – for the continuation of an allegedly "dying" ET race. But above all, they attempt to transfer culpability to the American government by claiming that "most" of the millions of so-called alien abductions in America can be attributed to covert experiments being conducted by the "military-industrial complex."[24]

---

themselves of influences from what many believe are actually the effects of alien abductions. There is, of course, no way of proving this, but it is interesting to note.

[24]   It was president Eisenhower who coined this phrase, using the occasion of his 1961 farewell address to the nation to baffle Americans by warning them about the increasing influence of a military/industrial alliance usurping power from both the executive and the people.

Googling "Rockefeller military industrial complex," will provide thousands of hits purporting to explain the link between the American Security Council and several major military weapons manufacturers, including Nelson Rockefeller's alleged influence.

The previous president, Harry Truman, made the decision to classify as top secret – the highest level of restriction that can

Perhaps this derives from some sort of utilitarian, or "collective" moral point of view which favors the greater good at the expense of the individual, but the primary focus of the new-paradigm camp – to the eternal irritation of the traditional, pro-science camp – is on the positive, healing, and spiritual benefits of contact and integration with aliens.

---

be attached to sensitive information – all things extra-terrestrial, and hand the problem over to the military. And although nobody suggests that he had anything but America's best interest in mind when he made this decision, Eisenhower, who inherited both the sworn-to-secrecy ET problem and the ASC, apparently felt just as powerless when it came to dealing with his own government as he was when faced with defeating the "alien menace," and, unable to disclose the problem without violating his oath, was obliged to issue a warning in code. Here's what he said:

> A vital element in keeping the peace is our military establishment. Our arms must be mighty, ready for instant action, so that no potential aggressor may be tempted to risk his own destruction...
>
> This conjunction of an immense military establishment and a large arms industry is new in the American experience. The total influence—economic, political, even spiritual—is felt in every city, every statehouse, every office of the federal government. We recognize the imperative need for this development. Yet we must not fail to comprehend its grave implications. Our toil, resources and livelihood are all involved; so is the very structure of our society. In the councils of government, we must guard against the acquisition of unwarranted influence, whether sought or unsought, by the military–industrial complex. The potential for the disastrous rise of misplaced power exists, and will persist. We must never let the weight of this combination endanger our liberties or democratic processes. We should take nothing for granted. Only an alert and knowledgeable citizenry can compel the proper meshing of the huge industrial and military machinery of defense with our peaceful methods and goals so that security and liberty may prosper together.

What, they demand, of the proliferation of difficult-to-deny reports of aliens who do not abduct, harvest, or threaten to genetically alter, control, and/or replace us, but appear to be genuinely benevolent and claim to be here solely for the purpose of enlightening us, and teaching humans to become better custodians of our planet?

Far from inspiring fear, their presence is said to be a comfort and even a "gift," which is actively sought out by millions of spiritually-inclined UFO religious cult members and many more independent believers in the good-ET message of universal cosmic peace.[25] New-paradigm UFO literature, which is proudly non-rational, is distinguished by its anti-science rhetoric, and dedicated to a combination of validating the supernatural and denouncing the "scientific/materialist" world view, and devising political schemes for getting the rest of society comfortable with the idea of aliens, so that we may all reap the spiritual benefits of integration.

---

[25] UFO religious cult members frequently perform rituals in which they actively seek to be abducted, offering themselves up to extra-terrestrials to use for whatever purpose they can serve. The success rate is believed to be approximately zero. At the same time, military research into extra-terrestrial phenomena has an important psychic component, and "psionics," or people predisposed to extra-sensory experiences, have apparently been successfully trained by them to summon extra-terrestrial craft. It is even claimed that these psychics can "control" such technology "with their minds." Spiritually-included UFO worshippers believe that this indicates the presence of the divine. It might also be that whoever is controlling the extra-terrestrial craft remotely, is leading humans to believe that they are telepathic. Or that telepathy can be explained scientifically, and that all humans have this ability, albeit to different degrees.

The new-paradigm, anti-science agenda, then, aims to promote ET spirituality by pitting it against science, and is based, it seems, on the popular belief that the two are mutually exclusive.

But whatever might be said of the stridently anti-science, "spiritual," position from a logical point of view, it nevertheless remains the case that it is almost certainly the proponents of this sort of movement who, if anyone can, will be able to effect the political and environmental changes necessary to prevent our species from destroying ourselves, our planet, and all the other life forms on it – and by extrapolation, that of higher-order ET entities and other worlds as well.

The only problem – apart, that is, from the ethical repugnance of rationalizing bad-ET behavior and the practical impossibility of anything remotely related to peaceful human-alien integration on our planet – is that its central premise would seem to be insupportable. For, even granting ET *influence* on both our planet and human evolution, is it really conceivable that the alien entities which we are observing here now are fully sentient, flesh-and-blood, living, dying, human-equivalent creatures possessed of human-compatible moral and aesthetic sensibilities and concerned primarily for *our* betterment? And having observed us since antiquity, could they really expect us to improve in any appreciable way?

Wouldn't it be more logical – or more practical, if logic is forbidden – to assume that the "aliens" increasingly being observed by us are manufactured tools of some sort, created by higher-order entities which cannot, for purely physical reasons, visit our earth in their natural form, but are, like us, essentially self-interested, and are parroting back and re-enforcing human religious precepts simply as the most efficient means of conveying the urgent message that our use of nuclear weapons resonates in their own habitat – *wherever* that is – and they will do everything in their power to prevent it, including terrifying us, threatening us, abducting us, and bribing us with advanced technology and the promise of everlasting spiritual rewards?

They might, as they claim, of course be true flesh-and-ET-blood inter-galactic missionaries, who have equally mortal lives, and are otherwise like us. But it also might be that they are 3-D manufactured bio-replica technoids of some sort, engineered and injected into our world for the singular purpose of convincing us to dismantle our nuclear weapons before we succeed in destroying their designers in a higher dimension.

Perhaps this higher dimension has no concept of religion at all. It might even be that, in spite of their advanced intelligence, its inhabitants are no more capable of understanding our behavior than we are of understanding the behavior of the cells which make up our own bodies.

Nevertheless, by observing us, they have been able to learn how our communication system *functions*, and have devised tools capable of mimicking human beliefs in an effort to control our behavior – in much the same way that we have created pharmaceutical tools in the effort to control the behavior of our own bodily processes at the micro level, without really understanding them.

The fact that aliens seem to be able to "shut down" a given individual's consciousness for a limited period of time, but not able to compel us all to behave according to their will, all the time, seems to reinforce this idea. The situation is analogous to humans being unable to control – or even to essentially *understand* – the complicated micro-processes which regulate their bodies, but capable of creating pills to do things like shut down specific pain receptors, for limited periods of time.

We can, in other words, create pharmaceutical tools to shut down the pain receptors in our brains, but it is impossible for us to ever essentially *comprehend* how our internal communication systems work because our micro-parts exist in a lower dimension, which is alien to us.

We can observe, indirectly, cellular interactions, and use tools to mimic and control them, but even when we can figure out how to do this successfully, what we are doing constitutes an indirect, one-way interaction devoid of any real understanding; we have simply, by a rote process of trial and error, hit upon something which "works." Higher-dimension entities, who are essentially unlike us, might be designing "extra-terrestrial" tools which work in precisely the same way.

An entity in a higher dimension might, through observation, have discovered that of all concepts in our belief system, religion seems the most successful in stirring humans into action. It notes that in the American portion of our system, Christianity dominates.

So it creates tools which mimic a Christian message as a means of convincing American humans to cease and desist in behaving in ways which are harmful to the entities in its own, higher, dimension, without essentially understanding Christianity, or indeed the concept of religion itself.

Note that if these religion-spouting extra-terrestrials were truly conscious, or truly capable of comprehending us in even the most elementary sort of way, they would have figured out immediately that such a tactic would never work.

The fact that this approach was even attempted in the first place as a means of controlling our behavior tends to confirm that such extra-terrestrial visitors are not inter-active, sentient beings at all, but rather pre-programmed intelligences – and this by entities so far removed from the psychologies and social structures which characterize our planet that they are inherently incapable of comprehending them in any meaningful way.

In the "containment" model of dimensions proposed here, while an inter-galactic alien would easily be able to decipher all sorts of subtle nuances of earthly language and behavior, it would be impossible for an extra-dimensional alien observer to comprehend the odd role that religious principles play here even in the lives of people who believe in them; higher-dimension ET would simply observe that a particular set of religious beliefs is espoused by the vast majority of individuals, recited weekly in ritualistic ceremonies, and evoked in virtually every aspect of society, from patriotic pledges to sporting contests.

He would note that public adherence to these principles seems to be especially paramount at the highest echelons of authority, and conclude that they are therefore the most powerful and effective principles to invoke when attempting to communicate with and control earthlings.

Like the micro-biologist intent upon finding the correct set of "instructions" to accomplish an effective form of cancer-inhibiting gene therapy, if he failed to communicate his religious message effectively the first time, ET would simply persist, altering its style and content until he "got it right" – for 70 years now, and counting.

Or, perhaps, until *we* got it right, and were able to appreciate the literal truth behind the "supernatural" events in religious texts which, interpreted properly, explain the universe as it really is, and man's place in it.

What else can he possibly do? Given the cosmically incomprehensible physical differences between worlds, higher dimensional extra-terrestrials would, it is suggested, be conceptually incapable of understanding us in any meaningful way. But even if they could, despite frightful abductions and displays of technological prowess, the indisputable "dimensional superiority," of extra-terrestrials visiting our planet does not afford them absolute power to control us, because it is just as impossible for them to violate the laws which govern their lower dimensions as it is for them to comprehend the beliefs and behavior of the lower-order entities like us, who inhabit them.

These extra-terrestrial visitors are, it would seem, limited to attempting to trick us into believing that although they possess unlimited telepathic and/or "supernatural" powers, because they also possess highly advanced moral sensibilities, they would never interfere with our freedom to do as we choose; they are simply attempting to influence us away from our path of destruction, and this *because* they are so good.

This does not, of course, account for the terror, intrusion, and killing perpetuated by some of their members. But morality itself might be a uniquely human concept, which higher-dimension extra-terrestrials neither have nor even need. The reality is that it is probably a mistake to assume that we can ever even attempt to truly observe – at least according to *our* understanding of what it means to "observe" – entities from other dimensions. It is equally problematic to speculate about the motives of higher-dimension entities, even if they are able to present themselves to us in corporeal forms which we are able to perceive, and initiate a (from our vantage point) astonishingly effective one-way communication with us.

What we do know, if the above model is even approximately accurate, is that simultaneous existence across dimensions entails that they cannot hurt us without injuring themselves. Another way of saying this is that we are all one being.

If this is beginning to sound more like Eastern mysticism than science, perhaps this is because, as the wisdom goes, "Truth is One," and all belief systems ultimately converge at some point. Possibly we have reached that point now. But that is a problem for theology, not logic. From a logical point of view, a proper description of ETs will require an appeal to an alternate scientific/metaphysical paradigm which can accommodate not only simultaneous existence across dimensions, but things like substance dualism, reverse causation, and most importantly, a new theory of evolution. In fact, we are badly in need of these things even in the absence of anything like ETs to explain.

Our understanding of dimensions needs to be expanded upward; single-substance materialism has never adequately addressed the traditional philosophical "mind/body problem," whereby minuscule electrical impulses in the brain are supposed to be capable of propelling (comparatively) massive human bodies into action; our conception of causation and the alleged "randomness" which is purported to rule the universe is incoherent; and the theory of natural selection is dated and incompatible with both the laws in the surrounding sciences, and all of the geological and (physical) anthropological evidence. So if it is possible to construct a new metaphysical paradigm which updates these old concepts, its ability to cohere with the sort of behavior anecdotally attributed to extra-terrestrials can be considered a bonus.

Because the inadequacy of the accepted theory of evolution would seem to underlie all of these problems, this paper offers an alternative. As it happens, the alternative proposed is also able to accommodate all of the above characteristics ascribed to extra-terrestrials. Briefly, it consists in the following:

1. Replacing the present ontological model of "one-substance" materialism with a two-substance dualism[26]

---

[26] It is crucial here to distinguish between what is traditionally referred to as "substance dualism," and the more

which posits an independent, immaterial element of some sort.

2. Adopting a "containment" model of dimensions which includes the ideas of permeability and simultaneous existence across dimensions.

3. Substituting the present "push" concept of contiguous causation with an end-state determinism, or "systems" model in which the component individuals are viewed as

---

modern concept of "property" dualism. Substance dualism, which is historically associated with the belief in souls, posits an immaterial substance completely separate from the body. This independence is what permits the soul to survive an individual's death. Property dualism, which views the immaterial element as "emergent" somehow, and existing only as part of a functioning human brain, has nothing to do with souls. Property dualism is designed to describe the non-physical properties associated with thought and consciousness in terms of an immaterial "mind" substance which is intimately connected to, and dependent upon the brain in some unknown way.

When philosophers attempt to solve the mind/body problem by an appeal to dualism, they are referring to emergent, brain-dependent, property dualism. What is proposed here is a full-fledged appeal to an independent substance dualism, minus the religious connotations. The idea is to revive the *concept* of an immaterial substance – not as something divine, but simply as another element which exists generally in nature, human consciousness being but one expression of it. Because of the historical connection between immaterial substances and souls, religiously-inclined individuals are often attracted to UFO reports because they read into them elements of the divine. But there is nothing inherently supernatural or transcendental about the immaterial; as a concept, it is neither more nor less divine than the concept of massless micro-particles, for example, or microwaves. It is simply another aspect of nature.

operating a collectively for the purpose of obtaining a pre-specified final state.

4. Abandoning the essential randomness of genetic recombination associated with biological evolution for an end-state, systems' model which allows for numerous causes, including the possibility of manipulation by entities in higher dimensions.

In the proposed "containment" theory of dimensions, an infinity of dimensions exists both "beneath" and "above" the macro physical world which we inhabit, and entities in all of the lower dimensions are simultaneously "contained within," and working co-operatively for the purpose of maintaining the thermodynamic functioning of the entity in which they are contained.

In this view, entities inhabiting the various dimensions are governed by physical laws exclusive to their specific dimension, but the barriers between dimensions are not impermeable. That is, it is possible for entities in higher dimensions to construct tools which permit them to observe and/or manipulate entities in lower dimensions, but not in ways which would violate the physical laws in the dimension being manipulated.

The different dimensions might also contain different sorts of materials. This refers not just to materials with different sorts of physical properties, but to substances which differ ontologically from physical substances: In other words, to various types of non-physical substances.

The concept of non-physical substances is not new, but because it is traditionally associated with the idea of immaterial souls and/or "minds," post-Darwinian scientists have tended to dismiss it as so much mystical nonsense. Analytical philosophers, who have also embraced a materialist, "one-substance" ontology, have done so at the expense of logic, because the mind/body problem – that is, the impossibility of minute physical/electrical brain events being able to somehow cause or "push" human bodies into action – cannot be overcome by banning the term "mind." It is argued here that the solution is to expand, rather than eliminate, the concept of immaterial substances. This project hinges crucially upon a re-definition of what is meant by the term "mind."

If the immaterial "mind substance" is unique to biological systems of (x) degree of complexity, and somehow emerges out of nowhere and implants itself in these organisms' brains at some unspecified point either during fetal development, or after birth, and then "dies" with the organism, then, as the materialist insists, the concept certainly is problematic. But what if the substance traditionally believed to be exclusive to human minds is in fact simply another basic element in nature? In this case, rather than being philosophically problematic, we would *expect* human bodies to be possessed of this property.

It might be that the ability to exploit the immaterial substance requires a considerable degree of cranial complexity – we would not, for example, expect plankton to be able to make use of it – but if the immaterial substance, whatever one chooses to call it, is generally available, and works in the same "non-physical" way that our own minds do, there is no reason why, given only the requisite degree of intelligence required to manipulate it, this element could not be incorporated into tools and technology in much the same way that we now use things like microwaves and X-Rays.

If extra-terrestrials have found a way to do this, then they would be able to "will" their mind-enabled machines into action in exactly the same way that we "will" our bodies into action. And we really do. There is no physical chain of causation there, despite what the materialist would like to believe. Brain waves do not "push" bodies along; only immaterial hopes, wishes, and desires can do that.

Whether humans are sufficiently intelligent to learn how to use the immaterial substance to communicate telepathically with others is an open question. Apparently, some people can already do this, albeit in a very restricted sort of way. But the possibility of communicating telepathically with machines, as our alien visitors are purported to do, seems remote, if not impossible.

If it were simply a matter of using technology as a tool to harness the power of the immaterial substance, in the way that we use various types of waves, for example, this would be feasible. But the fact that ETs are also able to turn off our "prehistoric" earthly machines at will suggests that the trick to using it is not to somehow "inject" an immaterial substance into our technology and then train our minds to communicate with it, but rather to train our minds to communicate with a substance which is already *there*. Moreover, it appears that their ability to manipulate objects telepathically is innate, and not something that can be learned.

Extra-terrestrials appear to be capable of using their thoughts to manipulate objects both naturally and effortlessly, in exactly the same way that we use our thoughts "telepathically" to direct our bodily actions. But it is entirely possible that their ability to do this is only a misperception on our part. That is, it might be that when ETs *appear* to be manipulating objects telepathically, they are in fact utilizing some physical aspect of their bodies which simply takes a form that we humans, with our correspondingly limited sensory capabilities, are unable to perceive.

It might even be that when higher-dimension extra-terrestrials set about to construct the physical forms which they would require in order to be perceived by humans in this lower, physically-different dimension, they purposely designed them to also include additional, dimension-specific physical properties which we are *unable* to perceive. In this way, they could make it appear that they were able to perform "supernatural" stunts like disabling our technology and transporting humans "telepathically," when in actuality they are simply utilizing aspects of their physical forms which we can't see.

With human abductees, who would be able to feel the physical effects of such manipulations even if they couldn't see the forms responsible for them, it is entirely possible that ETs are transporting them "telepathically" by utilizing technology from a distance.

Abductees who have been transported by unseen ET forces often speak of having the experience that their bodies were being "pulled" into spacecraft, exactly as though they were being sucked into an immensely powerful vacuum of some sort, for example. Perhaps they were. But what of abductees who claim to have been pulled through walls, or transported into spacecraft through the ceiling?

The only explanation which comes to mind is less of an explanation than another puzzle, but it is, apparently, a real phenomenon. This is the phenomenon of out-of-the body experiences. The idea here is that it is not their actual physical body which is being transported, but some non-physical aspect of their mind, or consciousness. It is, in other words, identical to the experience that people who have died briefly – during surgical operations, for example – report, of "seeing their bodies from above."

People experiencing intense trauma of both physical and emotional varieties also frequently speak of "leaving their bodies," and there are documented cases where people having such out-of-the-body episodes are able to describe events which were happening in various rooms that they "visited" when they "left their bodies behind."

People can also sometimes repeat, verbatim, conversations which took place in operating rooms, which they claim to have heard while "out of their bodies" under general anaesthesia. And note that this is not the same thing as patients waking unexpectedly during surgery; these reporters are completely unconscious and unresponsive, yet they can sometimes even describe the surgery itself, and claim to have watched it from "outside of their body" while they were sedated.

Our current one-substance, strictly materialist ontology makes it impossible to account for such experiences scientifically – at least within the presently-accepted scientific paradigm – but it's quite possible that there exists some logically-explicable, immaterial substance associated with consciousness.

Note that thoughts themselves are "immaterial," and there is little doubt that *they* exist. So possibly it is this same, immaterial, aspect of the person which is in fact being transported, while the reporter's physical aspect remains where it is.

To the abductee, this experience – whatever the explanation turns out to be from a scientific point of view – would be reported in terms identical to those used by the individual who has suffered great emotional or physical trauma, or has died briefly.

"'I' floated to the top of the room," the victim says, because it is this immutable immaterial consciousness, or awareness, which he identifies as *essentially* himself, while the body in which it is encased feels more like a vehicle of sorts, which he is obliged to use to cart it around – and which feels less and less like "himself," the older and more decrepit it becomes.

In the case of the abductee, the consciousness which the victim instinctively identifies as "himself" is similarly separated from his body, and separated against his will; the only difference is that the trauma which causes it to separate in the latter case comes from an extra-terrestrial source. In fact, there is a very real sense in which the latter scenario is even *more* reasonable than the former, for what, it must be asked, could be more capable of scaring the wits out of a body than an extra-terrestrial?

Whether ET abductions of this sort are performed in order to examine more closely the nature of human consciousness, or simply to terrify the abductee into warning the rest of the world that we had better do what they ask, because extra-terrestrials have the power to "pull people through walls," it's important not to dismiss such accounts out of hand, and brand the reporter a liar – or at best, deluded. Enough people have reported such experiences to make them a subject for serious scientific consideration, and these are sane, logical people who also don't believe that what they are reporting as having happened to them, is possible.

The fact that physical effects of these abductions are also often present does not necessarily prove that the comparatively mundane explanation suggested above is invalidated, or establish without a doubt that abductees' bodies are in fact pulled "supernaturally" through walls, violating every imaginable physical law and establishing the veracity of all things occult. For note that it is quite possible that while the abductee's consciousness has been separated from his body by extra-terrestrials – exactly as per in cases of purely terrestrial trauma – the body left behind is simply experimented upon where it is.

Again, note that the report in the case of abduction by extra-terrestrials seems to mirror exactly that of the surgical patient who "leaves his body;" the consciousness which the patient identifies as himself is transported somewhere else, while the body left behind is physically altered.

In other words, if it is possible for people to "leave their bodies against their will" when they are being emotionally or physically traumatized by terrestrial events, then why should it be considered impossible for them to leave their bodies against their will when they are being traumatized by extra-terrestrial events? The problem seems to be more one of our lack of scientific terms and concepts to explain out-of-the-body experiences in a general sort of way than a problem specifically related to verifying the existence of extra-terrestrial abduction events.[27]

---

[27]   This is a problem which is not limited to explanations of out-of-the-body experiences, but extends to scientific descriptions of virtually all human experiences. It extends to scientific descriptions of the physical world as well, and is generally regarded by scientists as a sign of success. There is even believed to exist some sort of "inverse rule" which decrees that the more inexplicable physical laws seem to mere humans, the more intellectually sophisticated they must be.

Possibly this has its roots in the desire to avoid anthropomorphism at all costs – which is typically equated with "objectivity" – but scientists routinely congratulate themselves about the fact that their theories are *inherently* incoherent from a conceptual point of view, and are only meaningful as mathematical expressions. This, from a philosophical point of view, also mirrors the ultimate mathematical nature of the universe – a concept which goes back at least as far as Aristotle.

But while it is certainly the case that our limited human perceptual abilities cannot possibly acquaint themselves with even a fraction of a fraction of what must be out there, it is also unquestionably the case that announcing – as scientists

At any rate, since visiting extra-terrestrials seem intent upon convincing us of their power to direct the course of human evolution, it might be that their astonishing displays of telepathic ability are intended to serve as a sort of enticement; that is: "Do as we command and we will endow humanity with similar powers." But unless – as is entirely likely – they have seriously misread us about this, too, it is difficult to imagine why extra-terrestrials would possibly believe that the promise of enabling us to invade the personal privacy of individuals in a species like ours would entice us to do anything.[28] And if they really *are* able to endow us with telepathic ability – or, at any rate, to teach us how to tap into some sort of free, telepathic feature which the universe already possesses, then their purpose can only be malevolent.

Making us fully telepathic would not appear to benefit extra-terrestrials in any way, as we are far too limited to

_____

repeatedly do – that the universe is fundamentally random, does not constitute an "explanation," let alone a scientific one, even if it can be construed in terms which are essentially mathematical. Indeed, viewed from a logical point of view, the statement reduces to a contradiction of sorts, because whatever is random is necessarily arbitrary, and not amenable to any sort of explanation at all.

It is also at least possible that some of our current inability to conceptualize laws said to express fundamental truths of nature arises from the acceptance of a set of metaphysical presuppositions, or "sub-scientific," beliefs which are both badly outdated, and entirely too simplistic to accommodate what we clearly *can* observe about both nature and human nature.

[28]   For a discussion regarding the concepts of consciousness and telepathy, and scientific research into the subject, see Appendix H.

be of any intellectual interest to them. And it would not serve either us or our planet very well either, even if, as they profess, this is their intention. On the contrary, the ability to use telepathy would probably only make us angrier, and more violent. The very idea of enhancing our level of awareness in such a way even between genetically close family members is so alarming – indeed it is abhorrent – that the possibility probably represents our greatest fear of an alien presence here on earth.

As a species, we seem to have an innate need for privacy, and the ability to invade one's thoughts constitutes the ultimate violation. It is difficult to imagine how a biological type which can barely survive an open mike would cope with such a situation except by resorting to mass executions. Note that we have burned "witches" for less.

The most reasonable explanation for ET displays of telepathic powers, then, is not that they are offering us the prospect of similar "enhanced communication," but that they desire to frighten us into compliance. Presumably, if they really *could* teach us how to invade each other's thoughts, they would have done it a long time ago – and gratefully observed us murder one another in rage, to extinction.

But regardless of how limited humans are – or wish to remain – in their ability to communicate with their neighbors, the proposal for a new metaphysical perspective offered here does specify a "collective" theory of categorization according to which our *species,* rather than each human, is viewed as the "individual unit."

This perspective does not deny that humans can be meaningfully described as individuals. It simply states that they can also meaningfully be described collectively – as inter-acting components of an individual species unit, or "system," which changes, or "evolves," over time. And that, knowingly or not, they work co-operatively to maintain the thermodynamic functioning of the greater system in which they are contained.

A systems', or "functional," approach to scientific explanation also entails an end, or goal-related, causal order. That is, instead of evoking the traditional "bump from behind" theory of explanation according to which events are caused entirely by some sort of contiguous association with prior events, the idea is that events can best be explained by reference to the end-states of the systems in which they are contained.

In the model proposed here, this functional causal order characterizes the relationship between systems and their components not only within dimensions, but between them. For want of a better word, the latter relationship will be referred to as the "containment" theory of dimensions. A useful analogy for understanding the functional/causal relationship between dimensions would be a description of the relationship between individual humans and the cells which comprise them:

Cells are "contained within" a human body, which constitutes a functioning thermodynamic system. But they exist in a lower dimension than the body itself, and are governed by a dimension-specific set of physical laws which humans do not – and cannot – fully understand.

We cannot communicate with our cells directly because they exist in a lower dimension, but we can "interact" with them by creating tools to observe and manipulate them in various ways. We also cannot explain their behavior – except by reference to their function in the human system which contains them.

To refer to, or "classify," a cell as an individual unit of some sort is certainly intelligible, but to speak of cells as *behaving* independently simply does not make sense; they can be understood only in terms of their function in the biological system as a whole, and their thermodynamic "purpose" in working together with the other components of the system to collectively maintain it. Because it takes place in a lower, micro dimension, individual cell behavior is not noticeable in our higher, "human," dimension. But if something goes radically wrong at that level, the results do register negatively at the organism level – sometimes to the extent of killing it altogether.

An example of this would be the sort of cellular malfunction which causes cancer: Our cells might exist in an unnoticeable lower dimension, but because they are inseparable from ourselves, we cannot kill large collections of malignant cells without harming the bodies in which they are contained.

And because the growth of cancer cells is pre-programmed by some set of illusive "instructions" which control cell growth, killing the cancer doesn't cure anything; the (faulty) programing will simply cause it to grow back.

If we want to cure cancer, killing it is only the first step; we will also have to "reprogram" the out-of-control cells in a way which will cause them to stop dividing in such an irregular, system-destroying manner. This involves first designing micro tools to observe and manipulate lower dimension entities, and, finally, learning how to "communicate" with them successfully enough to re-program their instructions.

To extrapolate, if individual humans are collectively contained within an entity in a higher dimension, that entity might be capable of creating tools to observe, manipulate, and even kill select individual humans, but it cannot harm us unduly without harming itself, because in an ontological sense, we are inseparable from it.

If we are pre-programmed to do destructive things which threaten the greater entity in which we are contained – such as destroy our habitats and nuke each other – the higher-order entity can eliminate isolated sections of destruction-causing individuals by selective killing, and even, in extreme cases, eradicating entire civilizations. But, ultimately, it cannot kill us all; it can only attempt to re-program us in a way which would cause us to stop killing ourselves.

To use the cancer analogy, perhaps in ancient times, when mankind also acquired the knowledge to create dimension-spanning weapons of destruction, thus damaging the health of the higher-dimension entity in which his habitat was contained, that entity attempted to "cure" the problem by wiping out entire civilizations, and returning selected collections of humans to the sort of "pre-civilized" conditions which still characterize isolated pockets of our globe – notably in places which were once culturally and technologically advanced.

But because, as a species, we seem to be pre-programmed for destruction, the stragglers who managed to survive eventually arranged themselves into similar civilizations, re-acquired the sort of knowledge and technology which humans are hardwired to understand, and used it for exactly the same purposes.

At some point, after successive, temporarily-successful annihilations, the higher-order entity decided that a more effective, long-term solution to the human problem might be to learn our communication system and attempt to alter our faulty "program" by re-instructing us towards a more peaceful path. Or possibly the "re-programming-by-persuasion" approach simply represents a first, therapeutically-conservative attempt, and if it doesn't work, eliminating the program completely by annihilation and a return to pre-civilization conditions follows.

In the absence of any concrete evidence, it is of course a matter of pure speculation. But if aliens did in fact visit our planet in antiquity, in many ways the idea that a faultily-programmed collection of humans represents a sort of "cancer" to the finite, higher-order entity which contains them makes for a more coherent model when it comes to explaining the historic relationship between humans and extra-terrestrials.

Contrast this with the generally accepted hypothesis that powerful, technologically advanced inter-galactic spacemen created humans by genetically manipulating the terrestrial animals which were present on earth, deliberately endowed us with our inherently destructive, species-specific characteristics, and then left us alone to kill each other in wars, keeping their existence a secret but re-visiting periodically to provide us with *more* advanced technology that we could use to destroy their habitats along with our own – while begging us not to.

In the model proposed here, humans, like cells, are part of a vast collection of cosmic components which collectively – if unknowingly – function in a way which tends to maintain the higher-dimension entity in which we are contained. This higher-dimension entity is sufficiently remote to be rendered oblivious to the workings of its individual micro-components as long as they perform smoothly. But if some errant micro-event radically injures or alters the natural order of the system by doing something like exploding a nuclear bomb, it will register in, and also harm the higher-order entity which contains it.

This would explain why extra-terrestrials are so interested in our nuclear technology. If they were from an "earth containing" higher dimension, then an atomic blast would injure them too, perhaps critically. It also explains why they do not – at least not very often – take any really drastic action, or simply annihilate earth altogether. Because simultaneous existence across dimensions renders us all ontologically inseparable, any serious damage higher-dimension entities inflicted upon us would also hurt themselves.

It would explain how they can "vanish" into thin air. If extra-terrestrials are tools created by entities in a higher dimension, when they disappear, they are simply being "pulled back in" to their own dimension, either to avoid detection, or to prevent us from destroying them.

And finally, it would explain how they can do things like communicate via telepathy and operate technology with their minds. These terms are used with considerable hesitation because they describe phenomena which, scientifically speaking, aren't supposed to exist. But what *is* supposed to exist is entirely a matter of speculation on our part, and necessarily specific to the scientific world-view which we happen to accept at any given time.

If extra-terrestrial entities originate in higher dimensions, we would expect the physical laws in their dimension of origin to differ from those in ours, just as we expect entities existing in the quantum realm to be bound by physical laws which are specific to lower dimensions. There is, in other words, no such thing as "*the* laws of physics."[29]

But even if there were, it would be ludicrous to presume that we have gotten them exactly right, in their entirety, at any given point in history.

---

[29] This might serve to explain the concept-defying nature of quantum mechanics. That micro processes are governed by physical laws which are just as "real" as the ones which apply in the macro word is beyond doubt, as scientists are able to both define these laws, and to apply them in technology. But from a purely conceptual point of view, they don't make any sense; their explication is coherent only mathematically. Typically, this is interpreted as a limit to human perceptual abilities. But perhaps the problem is an ontological one as well, and the different dimensions are characterized by physical laws which differ essentially, and cannot even in principle be reconciled.

Nevertheless, these things aren't completely fluid, and despite the obviously progressive nature of science, it also does not seem logical to assume that, given only time and effort – or even a generous injection of ET information – we could necessarily "learn" to behave in ways which are typical of entities in higher dimensions, especially if such behaviors entail the use of perceptual abilities which we biologically lack. Telepathy is a case in point.

For telepathy to work, thoughts would have to be non-private, publicly-observable objects which exist in some universal, immaterial substance, and are transmitted through this medium in a way which humans – typical humans, at least – lack the perceptual apparatus to perceive. Note that we are inherently blindered here; even if, in fact, all thoughts are perceptually available to all observers in every dimension, our limited human perceptual apparatus means that we can only access our own.

If higher-order entities possess an expanded perceptual apparatus which enables them to observe not only their own, but *all* the thoughts which are blinking on and off in their observational space, then higher-order entities observing us would be able to "read" (that is, to perceive) our thoughts, too – in a completely mundane sort of way which is typical to them.

They wouldn't necessarily be able to acquire any deep, dimension-specific, *understanding* of either us or our dimension-specific physical world, but by repeated observation, they might be able to "learn our language" –

in much the same way that Watson and Crick "cracked the (genetic) code."

Learning our language would enable a physically-incommensurable, higher-order entity to communicate with us by conveying information through the medium of the immaterial substance, then. But because in the model proposed here it would be empirically impossible for a higher-dimension entity to physically *enter* our dimension, if it wanted to communicate with us, it would need to create micro tools programmed to facilitate such communication.

And note that such communication would *have* to be accomplished telepathically, because vocalization requires a level of biological complexity that a tool – being as it is a non-biological machine – does not have.

So the apparently amazing ability to communicate telepathically, rather than providing evidence of advanced evolution, might instead be taken as another indication that the so-called "extra-terrestrial" is not alive. Or more precisely, the extra-terrestrial which contains us in a higher dimension is alive, but the tools that it sends down to observe and interact with us are just that: Tools.

As it turns out, it is quite possible to re-interpret the behavior of visiting extra-terrestrials from a polar opposite point of view, and describe exactly the same qualities that the ufologist assumes are "highly evolved," in terms of comparative alien *inferiority*. The situation is typical of theory-dependent reasoning, and resembles the famous psychology experiment in which a researcher checked himself into a psychiatric hospital in order to get an insider's perspective on how these institutions function, and, assuming that he was insane, all of the employees therein proceeded to interpret every single thing that he did in these terms.

The psychologist did not alter his usual behavior in any way, but conducted himself in an entirely ordinary fashion as he went about his usual business of acquiring information, intelligibly, rationally, and politely addressing all of the doctors and nurses – who knew who he was, and why he "claimed" to be there – and fraternizing with the other patients in a completely normal and unexceptional way.

The other patients, who were not informed of the new patient's identity or research project, immediately recognized that he was not insane, and insisted that he did not belong there – despite being themselves profoundly psychotic. But the doctors, all of whom had trained for over a decade in advanced university programs dedicated to the subject, reflexively assumed that he *was* insane, and proceeded to view everything that he did from this perspective. For example, when the researcher took notes of his observations, instead of inferring that he was a researcher taking notes of his observations, they wrote in their own notes, "Patient engages in writing behavior."

In the same way, perhaps ufologists, most of whom have a fanatical interest in both space, and science fiction, and have dedicated their lives to studying the subject of extra-terrestrials, reflexively view everything observed about them as "highly advanced" simply because of their prior theoretical beliefs.

In contrast, consider someone who has no interest in science, and no knowledge of things like the problems associated with the immense distances spacemen would have to traverse in order to get to earth, or the imagination-defying technology which would be required to make the trip possible, etc. He is confronted with an earth-visiting ET. What does he see? A small, angular, hard-crusted, awkwardly-moving entity which does not eat, sleep, speak, emote, or reproduce. *He* might quite reasonably conclude that it is some sort of mechanical device which can't do much of anything – especially compared to humans.

The fact that it can "disable a nuclear missile with its mind," it might be pointed out, is akin to turning off a switch, and really, what is so impressive about that? A *highly-evolved* entity, to someone with no prior sci-fi assumptions about the subject, would be able to do something like play the violin with its mind, or randomly generate new Shakespearean sonnets.

Again, this is all just speculation, because in the absence of any concrete evidence, we have only historical reports to go on, but it does illustrate how there can be infinitely many ways of describing the same phenomenon, and how it is possible to radically alter, simply by postulating a new theory, not only our understanding of the phenomenon, but our perceptions of it as well. It also suggests that regardless of which explanation for the phenomenon of ET telepathy is accepted in the end, the problem of whether the extra-terrestrials who have arrived here on earth are themselves highly evolved is a separate problem.

As for humans, it might be that even if an identical immaterial "mind" substance exists in both this and in higher dimensions, we are limited to mind-to-body, or – in exceptional cases – mind-to-other-mind interaction, whereas for more advanced entities originating in higher dimensions, mind-to-machine interaction is possible.

Viewing humans collectively, as inter-connected components in their containing species' system, and acknowledging the existence of an immaterial mind substance, in other words, might eventually enable at least some of us to learn how to communicate telepathically with *each other* – there is already considerable evidence to suggest that this is possible – but it doesn't follow from this that we will necessarily be able to communicate telepathically with machines,[30] even if the physical laws in our dimension don't forbid it.

It might be that it is humanly possible to control machines with our minds, given only the right sort of instruction, and the will to do so. But every species has its own natural limitations, and if the ability to communicate telepathically with machines falls outside the limits of our cognitive abilities, then, like flying, this is not something that we will ever be able to do.

---

[30]   It is important not to conflate "brains" with "minds" here, even though this is the convention. Brain-to-computer interaction, however it is managed, is a purely physical process which is being done now and does not require the existence of an immaterial "mind" substance. A telepathic mind-to-computer interaction would involve something like operating a mouse by concentrating really really hard, with no electromagnetic waves allowed.

What we *can* do, at least theoretically, is predict the future, since it is specified by the end-states of all systems in every dimension.

2: INTRODUCING A NEW PERSPECTIVE ON COSMIC EVOLUTION

This section argues that the course of biological evolution cannot be explained by an appeal to a single principle, like "survival of the fittest," suggesting instead that its causes are both numerous, and varied.

It offers a brief epistemological critique of the principle of natural selection, and explains how the over-all course of biological evolution might be subsumed under the same laws which govern the course of cosmic evolution.

It explores some of the metaphysical consequences of this new perspective, including the simultaneity of existence across dimensions; the possibility that entities from higher dimensions might have influenced the course of biological evolution on our planet; the reclassification of humans as an inter-connected collective with their species' system functioning as an individual unit, and the corresponding possibility of telepathic communication between components of this system.

Finally, an alternate, teleological, perspective on causation will be proposed as the ideal explanatory model for all physical laws, making possible the concept that the future is not only knowable, but the ultimate cause of the past.

Keywords: *Evolution, Entropy, Dimensions, Extra-terrestrials, Simultaneity, Causation, Teleology*

The following "post-terrestrial" perspective allows for a multiplicity of causes of evolution, including the natural influences of entropy on genetic recombination, one-time impact and near-impact events in space, and the possibility of interference by forces in different dimensions.

The proposed new perspective is based upon three essential points:

1. It dismisses the principle of natural selection as both unnecessary, and epistemologically unsound.
2. It proposes to subsume the laws of biological evolution under the more general laws of thermodynamics, which traditionally govern cosmic evolution only up to the point of the emergence of life, and;
3. It posits that these laws, rather than the laws of micro physics, are the most universal, or fundamental, of all physical laws.

In other words, it is the laws which tell us how things change through time, rather than the laws which tell us what things are comprised of, which ultimately both facilitate and ground our understanding of the physical world. These are the laws of thermodynamics, which govern all energy exchanges in nature.

Briefly, the laws of thermodynamics state that:

1. The energy in a closed system remains

constant;

2. Energy always flows from a higher to a lower state, towards equilibrium. This is the principle of "entropy," which calls for increasing disorder; and,

3. Entropy (or "disorder") increases in the most efficient, or probable, manner possible.

An example of this would be that when hot water (heat being a function of kinetic energy) cools, (kinetic energy slowing down) it does not first boil or freeze before becoming luke warm (its end, or "equilibrium" state), but arrives at this end state in the most "efficient," or probable, way, by cooling down gradually.[31]

---

[31] Before the identification of heat as a form of energy, caused by random movement at the molecular level, it was believed to be a sort of fluid, called "caloric," following the way in which it always flowed from a higher to a lower state. The identification of processes formalized in terms of absolute laws with what was subsequently believed to be the inherently random process of kinetic energy had the odd effect of rendering long-standing absolute laws of nature merely "probable." From an epistemological point of view, this is a serious discrepancy, because one would expect a process which was only probable to various degrees, to admit of exceptions, and these laws never do.

Philosophers attempt to get around this discrepancy by suggesting that because the laws are derived from inherently random processes, they cannot be *called* "absolute," but, for all intents and purposes, they can be treated as such, and applied confidently in engineering applications.

They are, in other words, "in principle" statistical laws, and therefore it is statistically possible for them to admit of exceptions, but the possibility is treated as more of an

The following describes a new method of applying these laws to the phenomenon of evolution. It also explores some of the philosophical consequences of viewing cosmic evolution in this new way, including the possibility of simultaneous existence across both higher and lower dimensions, the non-random or "end-related" nature of all scientific laws, and the corresponding idea that the future causes the past.

---

abstraction than a real possibility.

It seems evident that the discrepancy between the reality of these laws as absolute, and their epistemological designation as "probable," is a direct consequence of appealing to "backwards," or "push" causation to explain them. The substitution of forward-looking, or "pull" causation would enable us to describe them as absolute, as per the original assumption about these processes. That is, it would eliminate even the statistical, or "in principle" possibility that, for example, heated water in an isolated system could freeze before arriving at equilibrium.

3: THE THEORY

There exists a wealth of already-established general scientific laws capable of describing all relevant classes of events which fall under their respective rubrics, so it is preferable, whenever possible, to subsume any newly-observed phenomena under existing laws, rather than attempt to devise separate, specific, principles which lack both the generality and the "heft" that comes with laws which have stood the test of time. This is especially true of phenomena which, on the surface, appear to exhibit characteristics which divert from the usual behavior of events in their relevant classes. Evolution is a case in point.

The laws of thermodynamics are universally-applicable to all energy exchanges in nature, and as such are believed to govern the course of cosmic evolution in the usual way – up to, that is, the emergence of life. The evolution of organic systems, it is observed, goes against the law which stipulates that disorder increases, and exhibits instead a course of increasing *order*. To explain this deviation from the otherwise-universal nature of this class of events, evolutionists have postulated an "interfering principle" called "natural selection," whereby the course of biological evolution goes against the grain by first "selecting," and then perpetuating beneficial mutations.

The idea that there is a struggle which pits both the organisms and the species in a desperate competition to survive is so deeply entrenched that it is almost impossible to dislodge it, especially since this principle of "natural selection" also pits science against religion – having itself been both proposed and perpetuated as an alternative to creation.

But note that the principle of natural selection has not advanced in sophistication,[32] or progressed in any way since it was proposed, as one would expect of such a simple, "first-guess" hypothesis which is now well over 100 years old. Partly this is due to the logical structure of the theory, as it cannot, even in principle, be extended to accommodate any other classes of events.[33]

But the rationale for forbidding theorists to question either its relevance or its worth has not progressed either: Natural selection is still being upheld as the "only" alternative to creation. If for no other reason than this, it

---

[32]   The so-called "New Synthesis," which ostensively combined Mendel's genetics with Darwin's original principle, resulted in an astonishing degree of progress in the former, and no substantive change in the latter. There would, contrary to another deeply-entrenched belief, seem to be no "synthesis" between the two at all; Mendel's successors are mapping the human genome and cloning mammals, and Darwinian's are still out in the field, digging up old bones in essentially the same way that they did in the 19th Century, and attempting to make empirical inferences from the fossil record.

[33]   Although it does not have any other physical law applications, the principle of natural selection is frequently appealed to in the social sciences to ground theories in subjects like sociobiology, where it is used in explanations of human behavior and/or morality. It is also sometimes applied in philosophical fields concerning things like language, perception, and consciousness. This paper will not address any of its derivative theories in the humanities, except to point out that to the extent which their validity depends upon the cogency of natural selection, they are subject to the same criticisms.

would be well to subject it to a bit of critical appraisal to see whether there is something in the nature of the principle itself which makes it impossible to replace, and, if so, whether it ought rather to be rejected outright as unnecessary.

Ultimately, it is entirely possible that what we have been trained to think of as "evolution" isn't a single process governed by some simple, single principle at all. It seems more logical to assume that the forces governing change through time are complex and multi-faceted. There is no reason to believe that all extinctions happened for exactly the same reason, despite Darwin's attempt to reduce everything neatly to "the struggle for survival," so why should we assume that all new forms on the evolutionary continuum came about in exactly the same way? Some might have been caused by the natural force of entropy upon genetic information,[34] by which unimpeded change always tends towards greater disorder; others might have been caused by radiation-induced mutations resulting from specific, one-time cosmic events; still others might be the result of outside manipulation from entities in higher dimensions.

If science progresses as it is supposed to, eventually the very concept of "evolution" might fade from the lexicon altogether, going the way of alchemy and the luminiferous

---

[34] Information theory, which uses the principles of statistical thermodynamics, was developed and quantified in 1948 by Claude Shannon ("A Mathematical Theory of Communication") for use in the field of data transmission. The idea here is that genetic information, as a physical form of communication, falls under the same rubric.

aether as it is supplanted by entirely new concepts in this field of inquiry. But note that in its present form, the theory of evolution is impossible to either expand, or formalize; it is frozen forever in 19th century mode, with Darwinians defining "adaptations" after the fact; equating evolution with natural selection; advancing as an argument in its favor the claim that, as a scientific theory, it is better than the "creation myth," and, finally; crowing that it can "never be disproved."

A theory which can never be disproved would appear, contrary to the above declaration, to be suspiciously *un*scientific, as such ideas make any progress in the sciences impossible, even in principle. Discrediting the theory of natural selection, then, would seem to represent a critical first step in the advancement of the evolutionary sciences. This is not a difficult task.

As it turns out, natural selection fails to conform to even the most rudimentary of conditions required to make an explanation scientific, beginning with the problem of "unfalsifiability" stated above.

There might not be much agreement in philosophical circles as to what constitutes a "scientific" principle, but if there is one single thing which is almost unanimously upheld, it is the requirement of falsifiability. This means that it must be possible – at least in principle – to specify the conditions under which it might be proved false.

To put it another way, if a conjecture can't be proved false, then it is a religious belief, not a scientific one.

Natural selection, which tells us that the fittest survive,

cannot be proved false because it also defines "the fittest" as "the ones which survive," and operates on the assumption that even if the organism/species appears – as it frequently does – to be rather remarkably *un*fit, because its survival entails fitness of some sort or other, this fitness must consist in something that we just can't see.

Note how this explanation resembles various well-known religious principles regarding things like the inadequacy of man's intelligence when it comes to being able to grasp the mysterious nature of God's benevolence, or the literal truth contained in the scriptures. From an epistemological point of view, these things are regarded by their adherents as "absolute truths," and are presumed to be on par with mathematical truths in the sense that they are accepted with certainty. Like natural selection, they are "fundamental truths," and as such are not open to dispute; the only sense in which they differ is that they are not also advanced as scientific theories. Of course, they also are not observation-based, "factual," statements – but neither is natural selection, as will be demonstrated directly.

4: A CRITIQUE OF THE PRINCIPLE OF NATURAL SELECTION

If discrediting the theory of natural selection is a critical first step in the advancement of the evolutionary sciences, it will be useful to subject it to a quick test for falsifiability to see whether it is able to meet this basic requirement for an hypothesis to be scientific.

The result is that, while the principle of natural selection is not falsifiable, every single empirical claim advanced in *support* of the theory is false, period. Neither is this a new problem, which might, for example, have arisen only recently with the disclosure of some startling new evidence, and be amenable to a bit of an update, or fixed by the addition of some auxiliary principle. Rather, it is the same old problem which caused Darwin to hesitate for so long before he published his theory in the first place.

Darwin advanced his theory as an empirical inference from observation, conjecturing that because it was possible to observe distinct similarities between the organisms in different species, and species seemed to come and go in geological time, there might have been some mechanism whereby one species "turned" gradually into another.

Tiny random variations, he suggested, might, given the competition for scarce resources, result in the "struggle for survival," whereby the "fittest" emerged victorious and passed on their small advantages to their offspring – which themselves were able to out-reproduce the competition and pass on the advantage to their own offspring, and so on and so on, until eventually they became another species altogether.

Darwin resisted publishing his idea for an inordinate length of time, hesitating because, although the hypothesis sounded well enough in theory, it was contradicted by all of the available evidence; the fossil record, to be specific, was alarmingly lacking in any of the "intermediate forms" which would be required for one species to melt gradually into another. Eventually, he was pressured into publishing, and could but "look with confidence to the future, to young and rising naturalists" to find the missing forms.

Instead of finding the missing forms – or appearing, indeed, to be similarly discomfited by their absence – his successors simply ignored the problem, and accepted unconditionally Darwin's principle of natural selection. It was subsequently upheld as the only alternative to creation, all of the (considerable) empirical evidence that it was false[35] was suppressed, and the evolutionary paradigm perpetuated via a series of now-famous diagrams designed to illustrate what the "evolution of the species" *would* look like – if the species had in fact evolved in the manner which Darwin suggested.

The series of horses which range from the tiny mesohippus

---

[35] See Richard Thompson and Michael Cremo's "Forbidden Archaeology," (Torchlight Publishing, 1993) a 900+ page tome filled with examples. A more accessible, condensed version, called 'The Hidden History of the Human Race," (322 pages) catalogues hundreds upon hundreds of cases where evidence discrediting natural selection was suppressed, and the careers of the archaeologists who advanced it, destroyed.

of antiquity to the majestic, dressage-worthy equus of today? This is an imaginary example; there is no descent among any of them.

The various "ape-man to man" illustrations? More of the same. And this is not a mere question of time, or lack of resources: The probability that such lines of descent will *ever* be found is vanishingly small, because everything about the fossil record observed to date distinctly contradicts the notion of gradualism required of Darwin's theory, and displays instead species emerging seemingly out of nowhere, and then becoming suddenly extinct, for no apparent reason, with some remaining for immense amounts of time, and others simply vanishing. And far from following a single, observable pattern, the only thing which appears to distinguish any of this are the huge "gaps" between the various types of organisms which come and go in evolutionary time.

Part of the problem here is that the theory of natural selection was advanced as an empirical inference from observations – that is, from Darwin's observations about the morphological similarities between organisms in different species. And when a principle is advanced and upheld on purely empirical grounds, there is no clear rule for determining exactly how much factual evidence is required to maintain it, and how much, to discredit it.

If, on the other hand, a theory is stated boldly and unequivocally as "true," along with the specification of conditions which would render it false, then all that is required for its adherents to be forced to concede failure is evidence that these conditions have been satisfied. If one

were to state Darwin's theory in this form, it would look
something like this:

1. If Darwin's theory of natural selection is
    true, there will be "intermediate forms"
    between species.

2. There are no intermediate forms between
    species.

3. Therefore, Darwin's theory of natural
    selection is not true.

Even Darwin would have to concede this point. The logical
structure makes the conclusion not just highly probable,
but mathematically necessary. To defeat an argument like
this, one would have to demonstrate that it contains a false
premise. But the premises are both true. Darwin himself
expressed the belief that the truth of his theory depended
upon the discovery of such forms, and they have never
been found. In fact, every single aspect of his theory can
be subjected to the same treatment, with exactly the same
result:

1. There is no "gradualism."
2. Mutations are almost never beneficial.
3. Ill-adapted species prosper and reproduce.

If one states a theory by specifying the conditions under
which it would be proved false, all it takes is a single piece
of evidence to demonstrate that these conditions have been
met, and the truth of the theory is put into question.
Darwinians, conversely, argue for their position by
employing a form of reasoning which, instead of specifying
the conditions under which the theory would be false,

specifies the conditions under which it would be *true*. They then make a fallacious inference from the premises to support their conjecture. Here is a typical example:

1. If survival of the fittest is true, then the fittest species survive.
2. This species survived.
3. Therefore, this species is the fittest.

While this argument does have a certain surface appeal, it is patently fallacious.[36] The conclusion does not follow from the premises. But unfortunately, not all scientists are trained in logic, and the upholders of natural selection, following this fallacious line of reasoning, extend the definition of "fittest" to include properties which can't be observed but nevertheless *must* be there, because the theory demands it. This sort of reasoning is also patently fallacious, but the conclusion seems to follow from the first inference, and to people who believe in the theory of natural selection, it sounds right.[37]

---

[36] Technically, the fallacy is referred to as "*affirming the consequent*," or "*converse error*." It is similar in structure to the valid form referred to as "*affirming the antecedent*," or "*modus ponens*:"

1. If it is Tuesday, then we are in Belgium.
2. It is Tuesday.
3. Therefore, we are in Belgium.

[37] Darwinians employed this same sort of "anti-logic" when they classified the species. They did this by going to every corner of the globe and inquiring of the locals what sort of animals they had in their neighborhoods. When virtually all of the answers included a giant ape-man – called "Yeti," or "Sasquatch," or something along these lines – instead of taking the existence of such a creature as decisive evidence that their

If one's opponents reason fallaciously, it is pointless to argue with them. In fact, it is impossible. To argue productively, both parties have to follow the laws of logic. But there is nothing logical about any of the claims that Darwinians make, and if one accepts their model of explanation, they can never be refuted. Consider the following theory:

> "(Neo-)Darwinism stipulates that evolution is a slow, gradual process whereby small, random mutations produce enough beneficial changes to give the mutants an advantage over their non-mutant progenitors. The mutants then reproduce and continue to pass on their beneficial mutations until a new species emerges. There is a struggle for survival between the old and the new, and the new species wins, while the old dies out."

Every single one of the above statements is false, and every attempt to justify them involves fallacious reasoning:

1. There is not a shred of evidence to suggest that it is empirically possible for one species to "evolve" into another, and copious evidence to suggest that such a thing is impossible. At most, the fossil record is able to confirm only morphological similarities between species. And even with "artificial selection," although almost unlimited variation *within* a species seems obtainable, it never results in a new species.

---

theory of natural selection was false, they took their theory of natural selection as decisive evidence that Bigfoot could not exist, and refused to include the ubiquitous ape-man in their classification scheme. Then, just to be on the safe side, they made jokes about it.

2. Even if it were possible for one species to somehow "transform" itself into another by the passing on of random mutations, in nature, mutations are not "beneficial." According to the principle "We know that this is the fittest because it survived," there *must* be beneficial mutations fueling evolutionary change because there is evolutionary change. But this is a logically fallacious inference, called "affirming the consequent."

3. All of the available evidence suggests that most evolutionary change happens quickly, all at once, and for no apparent reason. But since the principle of natural selection stipulates gradualism, the Darwinian claims that either the intermediate forms are there, and just haven't been found, or says that dramatic changes can happen all at once if the individual is a "hopeful monster." The latter is not a *formal* fallacy, like affirming the consequent, but it does contradict both the letter and the spirit of the theory. It is also silly.

4. There appears to be no "struggle for survival" in the Darwinian sense that minor, or even tremendous changes in a species would give it a competitive advantage in a "war against extinction." Although it is impossible to say what might have caused extinctions in the distant past, what we can observe of extinction in known times is a process which might be better described as "the slaughter for supremacy," as completely defenseless, inherently vulnerable species are decimated by predators whose natural superiority could not be challenged even if, say, the passenger pigeon doubled in size, or the buffalo sprouted seven legs.

Darwin's theory of natural selection, then, has not progressed with time, and his confidence that future naturalists would some day find the evidence required to confirm it has proved entirely unwarranted. All attempts to justify the theory of natural selection employ either the use of logically fallacious inferences, appeals to "facts" which are contradicted by all of the available empirical evidence, or both. Yet the theory of natural selection is not only upheld, but is revered – and revered as a "triumph of science over superstition." How can this possibly be? The answer would seem to be rooted in two associated points, one of which is logical, and the other, sociological.

The first, and most basic, condition of a theory for it to be called "scientific," recall, is that it must specify the conditions under which it would prove to be false. But if one were to ask the Darwinian to specify the conditions under which his theory would be shown to be false, the response would be something like, "But natural selection is true! It can never be proved false!"

This, indeed, is taken by the adherent of natural selection to be the strongest proof that anyone can advance in favor of a scientific theory. He knows that science is supposed to be progressive, that older theories become obsolete and are set aside as they are outshone by newer, better theories. But to the Darwinian, there is something so fundamentally true, so *absolute,* about evolution that the theory can never, even in principle, be proved false or become obsolete. And how can one disqualify a theory if it can never, even in principle, be disproven?

For his part, the Darwinian simply refuses to take the

possibility seriously. This is the "sociological" part. There is virtually nothing of note in the literature which even attempts to refute any of the arguments advanced against the theory of natural selection. Regardless of the objection, the Darwinian's favorite method of response is to make jokes about religion, and insist that either you can accept natural selection, or you can believe that God created us out of dust in seven days.

This is an astonishing bit of reasoning, and it says something about the status of the evolutionary sciences that it is not only believed to be completely cogent, but actually upheld as the ultimate "clincher." But imagine how odd this sort of argument would sound if applied in any other field of inquiry, even in the 19th Century, when Darwin advanced his theory:

> "Fellow scientists and distinguished guests: Thank you for allowing me to present to you my theory pertaining to the nature of heat. Either it is caused by "kinetic energy," the random movement of excited molecules which slow gradually in the most probable manner toward the state of equilibrium, where, although still present in its original quantity, the energy becomes inert and unable to perform work – or God created steam engines and warms them up for us, because that's the kind of guy He is! Ha. Ha. Ha."

In addition to being unseemly, arrogant, and absurd,[38] the above option contains another logical fallacy, called a "false dichotomy." It means exactly what the name

---

[38] If this sounds rather like contemporary scientists making jokes about "little green men" when confronted with evidence of UFOs, that is probably because, historically, this is exactly what scientists do when presented with phenomena which they do not understand: Make jokes about it.

implies, and Darwin was responsible for originating the false "science vs. God" dichotomy when he advanced his theory of evolution as an alternative to creation. His anti-religious followers have been repeating it *ad nauseam* ever since, in order to ensure that the supremacy of natural selection can never be challenged, and sneering that *it*, in contrast, is "scientific."

In Darwin's defense, there *was* no other competing scientific theory of evolution with which to compare natural selection. But in this case, the correct approach would have been to simply present his theory as a reasonable explanation which accorded with both the facts of observation, and the theoretical laws in the surrounding sciences. Natural selection, as it happens, did neither, and Darwin himself was well aware of this.

It is, unfortunately, impossible to prevent people from pitting science against religion. But even if one were to agree to accept the *concept* of evolution on the grounds that the alternative – creation – is untenable, it still does not follow that *natural selection* is true.

To set the dichotomy up correctly in this case would be to state: "Either God created man in seven days, or humans arrived here in some other way."

If one rejects the former explanation, the later is still wide open. Within it is the possibility of an unlimited number of scientific explanations. Darwin proposed one called "evolution."[39] If he – or his contemporary disciple – wants

_____

[39]   Actually, he introduced the concept of survival of the fittest. The contemporary identification of evolution and natural selection is another problem, as the two concepts aren't

to argue for its legitimacy, the onus is on him to advance some sort of empirical evidence in its support; it is not sufficient to assert that one of the religious alternatives to it is the possibility of divine creation by a Judaeo/Christian god, and it's no good, so we win![40]

Another way of saying this is that the falsity of religion – if it is false – does not make necessary any *particular* theory of evolution; there are, in principle, an unlimited number of alternate theories of evolution. But if this is the case, where are the alternate theories? Why, in all of this time, has no-one ever advanced even one alternate explanation to the problem?

The standard answer to this question is that natural selection is just too good a theory to either question or compete with. Yet a closer look reveals another aspect to the theory which tends to make it impervious to competition: Natural selection creates the very problem that it purports to solve.

Because there is no one, single, cause of change through time, there can be no one, single, principle to explain the phenomenon, so it isn't surprising that nobody has ever been able to come up with a competing theory. The only really surprising thing is that "survival of the fittest" was ever accepted as an explanation in the first place.

_____

necessarily identical. In other words, there can be evolution without "natural selection." Not, of course, to the Darwinian, but only "logically speaking."

[40]   For a brief history of the profound cultural revolution which was the result of the general acceptance of Darwin's theory of evolution, and the continued science vs. religion schism, see Appendix G.

"Survival of the fittest" is simply a tautology telling us that because to survive, a thing must be fit enough to survive, if it survives it must be fit enough to survive. And while it is difficult to dispute the truth of a statement which is true by definition, this hardly qualifies for advancing it as an *explanation* of anything. Usually, this sort of inference is referred to as "circular reasoning." To *explain* something is to identify its *cause*.

But it seems clear that the course of change through time is multi-faceted; its causes are many, and therefore not amenable to description in terms of a single principle like "survival of the fittest," even if the principle proposed did not suffer from circularity.

Some evolutionary changes might be the result of sudden, isolated, impact events like the one which purportedly wiped out the dinosaurs. Others might be due to the natural forces of entropy upon genetic recombination, according to which genetic information becomes increasingly disorganized over time. Others might be connected somehow to geological phenomena which aren't well-understood, like the reversal of the earth's magnetic poles, and some might be caused by things like radiation fallout from near-collisions in space.

The theory advanced here takes a systems-theoretical approach, and entails the subsumption of so-called "random" genetic recombination (genetic instructions being a form of physical information) under the universal laws of thermodynamics, which govern all energy exchanges in nature.

An alternate interpretation of these laws makes it possible

to view organic evolution as a regular instance of the "slowing down" of cosmic energy, without any need for an interfering principle like natural selection at the biological level.

Any interference with the general law of entropy would, in this model, be due to the sort of cosmic and/or geological events described above, and not to regularly-scheduled, let alone predictable, phenomena. But we must also, according to this model, remain open to the possibility of interference from entities in other dimensions.

Whatever explanations we come up with in the end, it seems clear that the attempt to describe change through time by an appeal to the fossil record is entirely the wrong way to go about it. The fossil record is so incomplete as to render it essentially useless in this regard, even if making simple inferences from observation were not an unnecessarily dated technique for theory construction.

In the case of evolution, at any rate, it is possible to apply more sophisticated methods of explanation, such as appealing to established laws in the surroundings sciences, examining the process of genetic recombination from a new, information-based perspective, and incorporating what we already understand about the universe as a whole.

Instead of making inferences from dinosaurs, we ought, in other words, to regard the scanty bits of old bones still buried out there in the ground rather as the incomplete skeletons of an otherwise entirely theoretical perspective. Or relegate them to the museum where they belong, and just look at them and wonder, as we do at King Tut's tomb.[41]

But underlying all of this is the idea that the metaphysical underpinnings of contemporary scientific explanation might themselves be questioned and modified to better fit contemporary science. And to this end it is suggested here a new way of interpreting the laws of thermodynamics, which govern cosmic change in its entirety. By viewing these laws, rather than the laws of micro physics, as the most general, or fundamental, laws of the universe, it is possible to explain biological evolution as a non-novel instance in the course of cosmic evolution as a whole.

Our understanding of the world, in other words, ought to be based essentially not upon the micro-particles of matter which make it up, but upon the way in which this matter changes through time.

Because our understanding of the physical world depends crucially upon the idea that it is comprised of ever-smaller bits of micro matter, all of which follow their own unique laws in ever-deeper dimensions below the macrosphere that we inhabit, this would seem to be the logical place to begin any attempt to re-define the metaphysical assumptions underlying these laws.

Based, apparently, upon the assumption that the familiar physical world of unaided observation – or "macrosphere"

---

[41] A corollary to this would be removing and preserving all artifacts as they are found, instead of the industry standard (if it is still a standard) of recording and then putting them back into the ground for someone to dig up in the future. And to be destroyed by the natural forces of entropy.

– which we inhabit is the highest dimension, scientific inquiries involving dimensions are invariably aimed at observations of events which take place "beneath" it. That is, they are directed at splitting atoms and attempting to discover the microparticles which comprise them, the assumption being that these particles are identical to the particles which comprised the original matter of the universe. And this assumption, in its turn, entails the belief that it is an understanding of what the universe is ultimately *comprised* of, which will explain the way the world *works*.

From a cosmic point of view, it is believed that the various exotic particles which are released when atoms are blasted apart are identical to the original energy of the universe which was released after the big bang. As things cooled down, as per the law of entropy, the original energy of the universe started to "stick together" as it became frozen or solidified into clumps of matter – first, into the various subatomic micro particles, and eventually, into atoms.

The total amount of energy present remained constant, as per the conservation law which stipulates that the total amount of energy in an isolated system remains constant, but the energy was rendered "unusable" as soon as it became trapped inside the atoms. And it is still there, with all of its original force – as scientists demonstrate amply when they blast these atoms apart.

Such research is fine as far as it goes – indeed it is spectacular. But possibly because it is so spectacular, physicists have concentrated all of their energies on attempting to explain our macro world by observing its sub-atomic depths, and in so doing have internalized the belief that our own cosmos, or "macro" dimension, is the highest one. This assumption would seem to be decidedly anthropomorphic. It would also seem to seriously inhibit our understanding of evolution, because scientists believe – or contemporary evolutionary theory, at any rate, *implies* – that cosmic evolution somehow "stopped" when the cooling-down process reached the macro-physical state which characterizes our particular cosmic dimension.

This problem is apparently not addressed in the literature, but the present theory either entails, or presupposes, the belief that cosmic evolution simply "ended," for some inexplicable reason, when it arrived at its current state. The universe itself is said to be increasing in entropy, or "running down," as per the second law of thermodynamics, and it is also supposed to be expanding – and even accelerating – but its *structural* organization is presumed to be stuck in what humans perceive as its present form.

Once the "cooling down" feature of cosmic evolution reached the macro-physical state with which we are familiar, in other words, the energy in the universe continued to be depleted, as per the second law, but the "clumping-together-of-ever-bigger-bits-of-matter" part which produced our macro dimension in the first place, ceased. In other words, the universe has stopped evolving.

In contrast, the theory advanced here has it continuing to evolve structurally as well as increase in entropy, creating higher dimensions with correspondingly different, dimension-specific, forms. To demonstrate how this process might work, it will be useful to re-consider the assumption in terms of our ability to manipulate entities in lower dimensions.

It is generally understood that while the various micro-entities occupying dimensions below ours follow different laws, the dimensions themselves are not impermeable. That is, we can both "enter," so to speak, and interfere with, the goings-on in lower dimensions. We can even re-arrange and use the entities in them for our own purposes, over-riding the laws which would otherwise be in effect down there. Both physicists and biologists do this all the time. Physicists, for example, create electron tools in the sub-atomic dimension, which can be made use of in the higher dimension which we inhabit.

When a physicist lines up electrons to create a laser, he is manipulating entities in a lower dimension, where they are isolated and stuck together to serve his purposes, over-riding their regular behavior in that dimension. It is not possible to completely undermine the laws in this lower dimension – that is, it would presumably be impossible to make the electrons in a laser stick together if they lacked the natural ability to do this – but it is possible to manipulate them to do things which they would otherwise not naturally be inclined to do. So the scientist is, in a very real sense, not only "travelling" to a different dimension, but altering it to suit his own ends.

Biologists do the same thing when they manipulate organic entities in lower dimensions. We're all familiar with the process, but simply don't think of it in these terms. It began with selective breeding, which itself entails a micro-manipulation of sorts, and ends with engineering applications of genetics so sophisticated that they converge with science fiction.

Nothing about the above interpretation of dimensions seems particularly esoteric, as long as the interpretation is confined to a discussion of lower dimensions. But if there exists a veritable infinity of dimensions *beneath* ours, there is no logical reason to suppose that there is not also an infinity of dimensions above it.

And if there are – as seems logical – dimensions above us, then it also seems logical to assume that, whatever the entities in these dimensions might be like, they are almost certainly every bit as interested in tinkering with the entities in dimensions beneath them as we are. It is for this reason that the suggestion was made earlier that we should remain open to the possibility that at least some aspects of evolutionary change might be due to interference from outside forces from other dimensions. [42]

The latter idea is an amusing one, and might be left simply at that, but the former does tend to explain a few otherwise-inexplicable things about our own dimension.[43]

---

[42] The concept of infinity is used here for simplicity; since we have no way of knowing how many dimensions there are, it's the easiest thing to state, but by no means necessary in a theoretical sense.

[43] If our universe – or to be more precise, the macro dimension which is *referred to* as our universe – is contained within an entity in a higher dimension, and is not only expanding, but accelerating, then perhaps these two concepts are related in some way. The entity which contains it, after all, would not be inert, but rather subject to the same thermodynamic forces that our macro universe is.

But whether or not the containing dimension is relevant to its

It explains, for example, why we seem to be the only intelligent beings in the cosmos.[44] Perhaps this only seems improbable because we have under-estimated the vastness of the universe. In fact, it's entirely possible that everything in our little cosmos amounts simply to a tiny tinkering, and is created by the interference of forces far larger, in dimensions far higher than we can conceive. The "big bang" of our universe, in other words, might be nothing more than the smashed atom in another dimension.

It also explains how it is possible that our physical bodies can remain exactly the same (minus the increasing decrepitness caused by entropy, that is) while every single cell which comprises them dies and is replaced with

expansion/acceleration, the conjecture that our universe is contained within *something* at least renders conceptually coherent the idea that the entirety of what there is – or to be more precise, the entirety of what there is in the macro dimension which we inhabit – is capable of expansion. It might also serve to overcome the so-called "horizon problem," or the question of why vastly separated expanses of space have identical properties and temperatures.

[44]   This would also explain why the extra-terrestrial technology observed to date – which apparently comes from another dimension altogether – seems to violate "the laws" of physics; presumably in a higher dimension the laws of physics would differ in much the same way that they do in lower dimensions. Einstein's quest for a unified field theory – which is currently being carried on by cosmologists intent upon finding some mathematical formula capable of reconciling quantum laws with relativity – would, given this model of dimensions, be put to rest once and for all. The good news is that his "God doesn't play dice with the universe" would not only be vindicated, but issue in an entire new paradigm.

entirely new cells on an apparently regular basis.

Our bodies contain cells which operate according to their own, vastly different, laws, in a slightly lower dimension than ours. These cells are themselves comprised of atoms which are governed by laws unique to their own specific dimension, which are in their turn comprised of sub-atomic particles, and so on and so on down the line, to whatever happens to be the going smallest micro particle postulated. But because the smaller of these entities are "contained within" the larger ones, and all of them occupy different dimensions, all of them can be said to "exist simultaneously in every dimension."

Another way of saying this is that entities remain intact due to containing forces exerted by the dimensions above them, essentially oblivious to what is going on in their own container of dimensions beneath. Just as our cells die and are "replaced," then, we die and are replaced, completely unnoticed by the higher-order entity which contains *us*.

In other words, the entities in the higher dimensions are "holding together" the entities in the lower dimensions. If our cosmos itself is contained within an entity in a dimension higher than this macro sphere, then it is this entity which must be "holding the cosmos together."

The concept is so strange that the very vocabulary required to describe it is somewhat painfully lacking. But even still, there is a coherence here which makes a sort of sense, and does not, at any rate, positively defy the imagination. What does positively defy the imagination is the idea that all of the meticulously ordered, and constant, law-abiding activity in the various dimensions beneath us is completely random.

If one studies any branch of contemporary science, randomness seems to be at the root of it. In biology, "instructions" governing cell behavior are exactingly mapped out in advance and executed for the explicit purpose of keeping the organism running in the best possible order; the individual components of the system actually appear to co-operate and even sacrifice themselves for the greater good of the organism – and it is all purportedly "random."

The laws of physics are also purportedly based upon processes which are both ultimately and essentially random – despite the fact their consequences are absolutely, unwaveringly predictable via mathematically-precise laws which have stood the test of time, and never once failed. This can't possibly be correct.

The solution proposed here is to begin with the effects, or "consequences" of these so-called random processes, and explain their end states by reversing the causal order. A brief summary of the two different types of scientific explanation is all it takes to demonstrate how this might be done.

There are two types of scientific explanation:

1. Physical laws, which are sometimes called "descriptive," or "predictive" because they do both of these things (e.g., "If something is water, then it will freeze at 0 degrees Centigrade"), and;

2. The functional, or "teleological" laws associated with the biological sciences.

Teleological laws explain by making reference to functioning organic "systems," and describe how things work by specifying a particular end, or purpose – "*teleos*" being Greek for "goal."

An example of a teleological explanation would be something like "The heart pumps '*in order to*' distribute blood to the extremities," or "Sperm swim upstream '*in order to*' fertilize the egg."

Notice the difference in logical form between the two types of explanation: The descriptive law simply states what a thing will do; the teleological law tells us *why* the thing is doing what it does.

Traditionally, it was believed that the descriptive laws of physics were superior to, or intellectually more sophisticated than, the teleological laws found in the biological sciences, because the former appear to be both "value free" and contain no so-called "anthropomorphic" elements like references to "goals," or "purposes."   Many philosophers even believed that in time – or at least in principle – the biological laws could be re-formulated in purely descriptive terms, and that all scientific laws could ultimately be derived mathematically from the most fundamental laws of physics, creating a tidy, transitively-ordered, "unity of the sciences."

The idea of unitary science is of particular interest here because the study of evolution suggests a new way to make the concept viable. The "directionality" which defines the physical laws describing cosmic evolution makes possible a re-interpretation of causality which would also seem to eliminate the essential randomness underlying both physical and biological laws. By "re-interpretation," it is meant "reversal."

This is not, however, to suggest that the causal order be somehow mysteriously reversed; it is rather to suggest that the causal order is *already* reversed. All that is necessary is to approach the problem from a slightly different perspective, and this becomes evident. Indeed, it is already evident.

Reconsider: "The heart beats 'in order to' distribute blood to the extremities;" "The sperm travels upstream 'in order to' fertilize the egg;" "The bird feathers its nest 'in order to' protect its young." Etc., etc.

These principles all offer explanations for specific phenomena by describing the goal, or ultimate "end" of the activity in question. The heart, in other words, isn't just pumping away at random. There is an end goal there, and if it isn't obtained, the entire system falls apart. Another way of saying this is that the purpose, (or "end") determines (or "causes") the process itself. The end is in the future. The future, therefore, causes the past.

In physics, events are described in completely value-free, *teleos*-free terms. Reconsider: "If something is water, then it will freeze at 0 degrees C," for example. There is no reference to purposes, or goals, or integrated systems of any sort here; just the cold, hard facts. Again, it was also believed that as biology progresses, its laws will eventually become amenable to description in the same impartial, *teleos*-free terms.

There might still be some lingering dispute about this, but

presumably it is generally accepted that this is impossible, and biological explanations can never, even in principle, be "reduced, without remainder" or described in purely descriptive physical terms. There is something inherently teleological about biological processes which makes it necessary to appeal to the entire system in which they are contained, including its purpose, if we are to describe them at all.

"The heart beats because it is comprised of cells which are made up of atoms," in other words, would never constitute an "explanation," even if we knew everything that there is to know about every single micro particle making up the atom, right down to massless energy and infinity. The only way that it is possible to understand how the heart beats is to understand *why* it beats, and this involves not only *teleos*, or "purpose," but, ultimately, an explanation of the system in which it is contained.

It also means that there exist two, logically distinct, types of scientific laws to describe what is believed to be a continuum process – cosmic evolution – and this seems untenable. If there can be no non-arbitrary demarcation between organic and inorganic events in evolutionary time, why should there be two types of laws? Certainly this could be attributed to something like increasing complexity, but if it is possible to re-define these processes in a way which would facilitate descriptions in the same logical terms – at least in principle – then surely this would be preferable.

But even apart from the aim of unitary science, if the universalization of scientific laws itself is to remain a goal, then it is interesting to note that the logical structure of the most universal of physical laws, the second law of thermodynamics, is distinctly teleological – until, that is, the beginning of biological evolution, which "survival of the fittest" describes as being fuelled from behind by selective "beneficial" forces which are purported to behave in an overtly non-teleological sort of way. If it is recalled that evolution did not begin with the emergence of organic systems, but with the big bang itself, and represents the most basic process in the universe, this is an astonishing discrepancy, both from an empirical, and an epistemological point of view.

The laws of thermodynamics govern cosmic evolutionary processes, whereby the original energy of the universe "cools down" into ever-increasing lumps of matter, and describes their end state as thermodynamic equilibrium. When organic processes emerge, the law of natural selection is believed to take over, and cause subsequent evolutionary events to positively go *against* the universal second law of thermodynamics, which stipulates increasing disorganization, and tend in the opposite direction, toward ever-increasing *complexity*. But this cannot possibly be correct, even if it can be established that the total amount of the energy in the system remains constant.

In the first place, energy is supposed to be depleted in the most probable manner possible. The principle of natural selection essentially guarantees that the opposite is the case. But more importantly, it just isn't possible for any natural process to "go against" a fundamental principle like entropy; this would require some sort of interference from outside of the system, and there *is* nothing "outside the system" of the earth and its atmosphere. Yet the complexity of the organisms is clearly *increasing*, rather than decreasing – as stipulated by the law of entropy – in evolutionary time.

The solution proposed here is both simple, and obvious. To show how the course of organic evolution might fall under the rubric of entropy in the usual way, with no "interfering" principle like natural selection required, it is necessary only to re-interpret the concepts of "order" and "disorder" slightly. And as these concepts are used in the evolutionary sciences, where the process is not expressed mathematically, this can be done simply by making the definitions of the terms clearer.

The second law of thermodynamics stipulates that disorder increases toward equilibrium. "Disorder" in the evolutionary sciences is defined in an everyday, "disorganization" sort of way. This being the case, it is possible to simply interpret the original free energy of the universe as being "highly organized" in the sense that it was neatly divided into single, separate, points.

Organic systems, in contrast, are highly *disorganized* at the genetic level. And they increase in disorganization as genetic information is replicated and duplicated and divided and perpetuated or lost forever all down the evolutionary line. The result? Ever more complex organisms at the macro level – complete with all manner of genetic disarray, dangling micro bits, and deleterious mutations. In other words: *Disorder.*

Moreover, this coheres perfectly with the law which stipulates that disorder increases in the most efficient manner possible: Note that the evolution of organic matter itself has resulted in biological systems which are far more adept at depleting available energy than is solid, inert matter, so the system in which they are contained is also running down nicely – or at any rate it is running down exactly as per stipulated by the law of entropy.

Subsuming biological evolution directly under the laws of thermodynamics, then, would seem to be easy enough. But the really interesting thing about all of this is that when one examines the logical structure of the laws of thermodynamics, they – unlike any other physical laws – appear to be distinctly teleological. In fact, they are expressed in precisely the form that is supposed to distinguish laws in the biological sciences, and even specify an end goal: Thermodynamic equilibrium. And since they also describe how micro-entities evolve through time into solid matter, they appear to be even more fundamental than micro physics. They might, therefore, give us a new logical model for the laws of physics in their entirety.

Another way of saying this is that if the laws of cosmic evolution are teleological, and the laws of cosmic evolution are the most fundamental laws of physics, then all physical processes must, ultimately, be teleological.

So instead of viewing teleological laws as "pre-scientific," and attempting to reduce them to non-teleological principles, we should be doing this the other way around, and trying to reduce the purely descriptive – and entirely static – traditional laws of physics to the interactive, directional, end-state teleological laws of cosmic evolution. This would, at least in principle, render all the laws of science logically compatible. It would also eliminate the randomness which is now believed to underlie the processes which these laws describe.

A teleological explanation appeals to the purpose, or "end" of a particular process. It tells us that something happens "in order to," or "for the purpose of." Such processes cannot be random, because they explicitly state that something happens in order to fulfill a specific purpose. The alleged "randomness" pops up only when one reasons backwards, and attempts to isolate inherently connected events by reducing them to simple descriptions of their components.

The universe itself is an inter-acting, evolving system. It can therefore only be described in systems' terms, and systems are never random. The components of a system operate according to some pre-determined purpose or end; their entire reason for behaving in the way that they do is always directed at some future state. As components of this system, all physical and biological events, then, must also be "caused" by the same, pre-determined, future state.

And what has any of this to do with dimensions? Just this: If one takes as a model the "system" of a physical body and its components, and considers the way in which cell behavior in its lower dimensions always tends towards the goal of "propping it up," and then assumes that there are dimensions higher than the one which contains the individual bodies here in our own dimension, it stands to reason that whatever the individual bodies in our own dimension do, is done collectively, for the purpose of propping up the entity which contains *us* in a higher dimension.

Individual humans, in other words, can, at least in principle, be explained scientifically as functioning, or inter-acting, components of a larger system – which is, like all systems, inexplicable apart from reference to its function, or pre-determined end state. Like the internal goings-on within our own bodies, according to which our micro-components are co-operating in order to maintain our over-all thermodynamic function in the most efficient manner possible, then, we must be (unknowingly) co-operating to maintain the entity in the larger system – or dimension – in which we are collectively contained.

What we might eventually come to know about entities in the extreme upper dimensions is a subject of pure speculation, but because it has been suggested that at least some of the extra-terrestrials purported to have visited our planet are arriving from a higher dimension much closer to home, this would be an excellent time to begin.

The first thing to note is that *direct* contact – that is, physical contact in their usual, dimension-specific material form – from higher-order aliens of any dimension would seem to be impossible, even with teleportation. But this certainly does not preclude their ability to down-visit indirectly, via specially designed micro tools of some sort. As for "visiting up," the situation would seem to be a bit more complicated.

It might be impossible for lower-order entities to ever visit the dimensions above them, even indirectly. Indeed, the idea itself is virtually incoherent: Does it make sense, for example, to speak of our cells "coming up" to visit *us*? Simultaneous existence across dimensions seems to entail not only that they are already here, but that *because* they are already here, they can't come up. Again, the requisite vocabulary is sadly lacking, but the concept seems clear enough.

To get a better sense of how simultaneous existence across dimensions works, it might be useful to compare it to the concept of alternate dimensions, as they are typically portrayed in science fiction.

In science fiction films there is always some sort of portal, or "gate" which conveniently opens into the imaginary alternate dimension. And although it is sometimes invisible, or difficult to access for some other reason, once they do find it, the characters just walk right in, using their regular movie star bodies. This is called a "parallel dimension."

What they find in these parallel dimensions is of course fantastic, and the creatures they encounter are often frightening and terribly unattractive, but the entities inhabiting parallel dimensions are – even when they can do things like shape-shift or become invisible – nevertheless wholly macro-physical in all of the same essential ways that we are. Indeed, there is an entire genre of science fiction films dedicated to portraying parallel dimensions as so similar to this one that the only detectable differences relate to a particular character who is classified as "crazy" when he has the misfortune of slipping in and out of them involuntarily.

The laws of causation in parallel dimensions also work in the same, "contiguous-bump-from-behind" sort of way which characterizes traditional explanations in physics, so positing them would seem to fall squarely within the present scientific paradigm – where the existence of alternate dimensions would, however entertaining, be ultimately just as random and inexplicable as our own. This is not to suggest that parallel dimensions do not, or cannot, exist, but only to point out that fictional portrayals of extra-dimensional travel do not represent dimensions in the same metaphysical sense imagined here.

In the metaphysical sense imagined here, simultaneous existence across dimensions entails dimension-specific physical forms which both differ essentially, and already occupy, all of the different dimensions. Upper-dimension entities, which "contain" those in the lower dimensions, can construct micro-tools which enable them to indirectly observe and manipulate entities in the lower dimensions. It is only in this sense that they can be said to "travel down" to lower dimensions. But because of the containment feature which characterizes simultaneity across dimensions, the only viable method for "travelling up" would involve exploiting some universal, immaterial, element common to all of the dimensions.

If we are "contained within" a higher-order entity in a different dimension, it doesn't make sense to speak of going up to visit it, because in an ontological sense, we are already there. What does make sense is to speak of inter-acting, or communicating with it *across* the dimensions.

Presumably, higher-order entities will be aware of our existence in much the same way that we are aware of the cells contained in our own bodies. When we observe cellular behavior, we see that cells are following their own dimension-specific laws in a pre-determined sort of way which looks to us like a "communication system," but can't possibly be – at least not in the way that humans understand communication. That is, although their behavior makes them appear to us as if they were aware of each other, the very idea that they might be is absurdly anthropomorphic. And if they lack the complexity for a consciousness which would enable them to even be aware of each other, how could they possibly be aware of *us*?

The contention that cells are indeed "aware" of each other, and communicating effectively – in some dimension-specific way[45] – can certainly be made true by evoking highly stipulative definitions of the terms in question.[46] We might, in other words, meaningfully state that cells are both aware of, and able to communicate with each other, "in their own, cell-specific, sort of way," despite the fact that they lack the consciousness required to communicate in the way that *we* think of as communication. But saying this does not add anything to what we already know about cell behavior, or contribute to metaphysics in a general sort of way. All it does is suggest that although cells

---

[45]   It is also sometimes suggested that everything living has "agency," as an inherent property of life. The opposite is argued here, but the underlying concept is identical: The idea of autonomy is incoherent if only some organisms possess it.

[46]   For a discussion on freedom, determinism, and consciousness, see Appendix I.

cannot communicate with *us*, we can, if we like, still refer to cell-to-cell interaction as "communication." Or possibly "communication without consciousness."[47] But it is not their lack of consciousness which is preventing cells from being able to communicate with higher order entities like us.

Obviously, entities which lack the complexity necessary for perception are not going to be able to perceive higher dimensions, even if they do possess the immaterial element which makes consciousness possible. But note that humans are incredibly complex, and this does not enable *us* to have awareness of higher dimensions; we are "cut off" from them in exactly the same way that our cells are cut off from us. And this "dimensional divide" can't be a physical barrier, because all entities are physically contained within the same system of the universe. There are no portals to open here; the divide is a purely perceptual one.

Whether or not lower-order entities can be said to perceive each other in any meaningful, dimension-specific, sense of that term, the fact remains that for all of their complexity, humans also lack the perceptual apparatus to observe, let alone communicate with, entities in higher dimensions. And this might well be a universal principle, which applies relatively to all lower-order entities across the dimensions.

---

[47]   Note that there is a big difference between "communication" and "information processing." The processing/dissemination of information does not presuppose consciousness or intentionality or anything else traditionally associated with human consciousness.

We can observe – albeit indirectly – entities in the lower dimensions, manipulate, and even learn to communicate with them in a limited sort of way. But our comparative lack of complexity entails that, like cells, we also lack the perceptual apparatus to observe "upward." In other words, if we want to interact with entities in higher dimensions, they will have to come to us, in a form that we *can* perceive. And if we want to communicate meaningfully with these forms, they will have to find a medium which is common to us both.

If UFO reports are to be believed, the first condition has already been met. As for the second one, since the immaterial substance which permits us to be conscious is presumably also used by entities in higher dimensions in the same way, even though it could obviously not permit us to *reason* at their level, it might permit them to communicate with us if they sent the appropriate micro tools down to visit – in a "telepathic" way which they would find astonishingly limited, and we would find extraordinary.

CONCLUSION

Humans might (or might not) have untapped abilities to communicate with higher-dimension entities. The only thing which seems certain is that the principle of simultaneous existence across dimensions means that we are already there – everywhere. But since the perceptual apparatus of entities in any given dimension is limited by their dimension-specific physical form, any meaningful inter-action between dimensions would probably have to involve a shift of consciousness – whereby "consciousness" represents but one expression of a universal immaterial substance common to all dimensions.

Another way of saying this is that the universe consists ultimately of a single, infinite, entity, with a corresponding infinity of inter-connected, "contained-within-it" individual components, functioning co-operatively in an infinity of dimensions in order to maintain the thermodynamic functioning of the system as a whole. It might also be comprised of an infinity of different substances, both material and immaterial. But, if only for the sake of simplicity, it would seem sufficient to posit just two: The material and the immaterial.

5: SOME PHILOSOPHICAL IMPLICATIONS

The concept of randomness is closely connected with the problem of free will and determinism, or the dilemma that whether or not we live in a deterministic universe, man has no freedom, or "agency."

The problem is posed in terms of a sound deductive argument whereby if you accept the truth of the premises, the conclusion, which follows mathematically, also has to be true. It goes like this:

1. Either events are determined, or they are not-determined.

2. If they are determined, then man is not free.

3. If they are not-determined, then they are random.

4. If they are random, then they are not free.

5. Therefore, man is not free.

This is a logically air-tight argument, so unless it can be proved that one of the premises is false, we are obliged to accept the conclusion, like it or not.

Philosophers who believe in free will tend to attack the second premise, and suggest that man's actions might be determined and still be free if they are determined, or "caused" by freely-chosen beliefs and desires. This, unfortunately, only pushes the problem to a lower level, and it becomes one of establishing how the thoughts associated with beliefs and desires can be free – which is doubly difficult because there is no definitive description of what thoughts *are*. All that is really known is that they seem to be immaterial in some essential way, which, if they are supposed to be responsible for "causing" material, or physical events like human actions, leads to what is called the "mind/body problem."

The mind/body problem has to do with the difficulties involved in attempting to establish a cause-and-effect relationship between the immaterial (thoughts, desires) and the material (physical actions).

Emotions, desires, beliefs, etc., are "immaterial" in the sense that they do not occupy physical space. But for one event to "cause" another, there is supposed to be some sort of physical contact, or "push" to get things going. And clearly, immaterial forces (thoughts) cannot move, or "cause," physical bodies to do anything in this way. There is, therefore, an "ontological divide" between the material and the immaterial which would seem to make it impossible for thoughts to be the cause of human actions.

Typically, philosophers attempt to bridge this gap by an appeal to materialism – the idea that there is only one sort of ontological material in the universe: physical matter. If there is no such thing as immaterial matter, then there is no problem; thoughts and desires are simply equivalent to physical brain firings of some sort.

The trouble with this solution is that it does not appear to be possible to establish an identity between emotions, desires, etc., and their associated neural firings, or "brain events." The two types of events are not equivalent at all. For although their associated brain activity is easy enough to identify and record, the thoughts *themselves* lack what is referred to as "public observability" – that is, they are, in essence, subjective and purely private experiences of some sort or other. We know this much, even if we don't know exactly what they are, or what sort of ontological form they take.

But even if it were possible to establish a one-to-one identity between thoughts and brain firings, it is still impossible to imagine how neural firings could "cause" our bodies to move in the robust way which is required of causation. Bodies are large. Little micro bits of electrical activity would seem not to be sufficient to push them forward into action.

There is something missing in this equation, and the proposed solution involves switching the focus away from concentrating on the attempt to establish the physical nature of thoughts, towards the problems associated with the concept of causation itself. The first thing to note is the considerable difference between being "determined," and being "pre-determined."

For the universe to be coherent, some sort of determinism is required; that is, events must be "caused." If they were not caused, then everything would be random, or arbitrary, and not amenable to predictive description or indeed comprehensible at all. Generally, this is taken to mean that events must be caused by something pushing them along from behind. But what if, instead of being pushed from behind, events were caused rather by a determination *forward*?

Another way of saying this is that perhaps a *pre-*determined end-state of some sort, rather than a push from behind, is ultimately responsible for the orderly, and obviously determined, nature of physical events.

This "forward," or "pull," sort of causation is distinguished by viewing the cause as being "ahead of," rather than coming from behind, as it does in simple determinism, and it seems much better able to explain the relationship between thoughts and behavior. In fact, it is possible that the mind/body problem itself is the result of the "backwards" reasoning associated with determinism in the first place, and that the attempt to identify thoughts with brain activity actually exacerbates, rather than solves the problem.

Even if, in other words, it were possible to establish a one-to-one identity between brain events and thoughts, *and* to somehow establish that tiny electrical neural firings could "push" huge human bodies forward, this still would not demonstrate that human actions are free. All it would do is lead to the question of how it might be possible to characterize the electrical activity associated with brain events as not only "free," but capable of the sort of thought which is believed to compel bodies into action.

Electrical activity operates according to its own laws in a lower dimension, and would seem to have nothing whatsoever to do with human desires and aspirations. And if, in its turn, we try to explain electrical activity via this model of causal explanation, we would have to go even farther back, and try to understand the source of whatever micro-physical entities are ultimately involved in neural firings – which would seem to be even *less* inclined to exhibit the qualities required to explain how human action is free. The farther back you push this "explanation," then, the less it seems to have to do with human activity at all, let alone human desires and aspirations.[48]

The problem, it is suggested, has to do with the "push" itself. If, instead of describing human behavior as being "pushed from behind" by various micro/electrical processes in the brain, we described the human individual as an inter-connected "system," complete with specified, abstract, end-states which operate in exactly the same way that abstract, pre-determined end states operate in any other type of system, it would be possible to characterize human behavior as processes *tending toward* its specified

---

[48]   Another philosophical solution/description of the relationship between mind events and brain events is called "epiphenomalism." The term "epiphenomena" is also frequently used in the UFO literature, in a highly stipulative – or at any rate subject-specific – way, but its precise meaning is impossible to determine from the context.

In ufology, it seems to refer to the immaterial mind substance which is used by extra-terrestrials to telepathically communicate, levitate objects, and otherwise perform actions at a distance. However, in traditional metaphysics, where the term derives, the epiphenomena can do none of these things.

Philosophers who believe in an immaterial mind substance which lacks causal efficacy are referred to as "epiphenomenalists" (literally, "above the phenomena"). Thoughts, in this model, are immaterial in the standard sense that they do not occupy space, and are also "private" in the sense that they are not publicly observable. But, unlike the immaterial minds in substance and property dualism, which are purported to be the cause of all human actions, the epiphenomena are not believed to be the cause of anything; rather, they represent the "offshoots" of brain activity, and simply float "above the fray" whenever the brain fires.

end-states – which are, of course, the goals, aspirations, duties, and, all too often, the unreasonable demands of others.

Colloquially speaking, in other words, instead of being pushed from behind by unfeeling micro-electrical processes in their brains, humans are being pulled into the future by their desires.

The above proposal involves dispensing with the concept of "push" causation – possibly in its entirety – and this is not a new project. The idea that there is no special force called "causation" derives from the empiricist philosopher David Hume, (1711-1776) who pointed out that when we see two types of objects standing in a relation we refer to as "cause and effect," all that it is possible to observe are the objects themselves; there is no force, or "added feature" called "causation" which makes the so-called causal relationship necessary. What could it possibly be? Some magic, invisible power?

This point is especially relevant here because if the "necessity" of the causal relation is impossible to establish, then the "before this, therefore because of this" *order* of the causal sequence may also be called into question.

It is only our peculiar observational vantage point which makes us interpret the relationship in this way; from a logical point of view, when one observes that two types of events always go together, there can be no reason to suppose that it is the former event which "causes," or "makes necessary" the latter – we might just as well conclude that it works the other way around. Take away the "necessity," in other words, and the necessity of *order* goes with it.

According to Hume, when (x) is said to cause (y), we infer that the relation of cause and effect is somehow "necessary" simply because we have seen the phenomenon in the past, and have become accustomed to observing that events of the same type seem always to go together. Hume referred to this relationship as "constant conjunction," and insisted that there can be nothing in the relationship beyond the conjunction itself.

That is, there is no other object, or force, or will, that *requires* the event which occurs first to "make necessary" the subsequently-observed event. The only requirement is that the two make some sort of physical contact, or, as he expressed it, be "contiguous."

It was the alarming absence of this sort of physical contact, or "bump from behind" which led so many scientists to reject, and even to ridicule Newton's theory of gravity, which postulated "action at a distance," by describing a force capable of acting on objects via what was at that time referred to as an "occult" power. By "occult," it is meant that there is some immaterial, or "supernatural" force at play – which is of course strictly forbidden by science.

It is this same abhorrence of occult powers which impels contemporary philosophers of science to reject the concept of causation, which they, following Hume, dismiss as "mystical" and "unscientific," and prefer to speak instead in terms of "functions."

"Heat," they might say, for example, "is a 'function of' kinetic energy."[49]

It might look as though one term is simply being substituted for another here, but the meanings of the two terms are actually quite different, even if the results are identical. A big part of the project proposed here is to bring causation back, in a full-fledged, "new occult" sort of way which derives from the idea that there is nothing necessary about the "before this, therefore *because* of this" relation.

Philosophers in the field of metaphysics, who typically neither know nor care about things like "functions" in science, but discuss and argue about causation in the traditional way, refer to the two different types of causation as "push" and "pull" causation, and are intrigued by the latter concept because it would seem to entail the same sort of abstract, "occult" forces that gravity did when Newton proposed his now-famous law.

Following Hume – as, it seems, one must – they concede all of the problems associated with push causation, and can find no logical means of undermining his critique and re-establishing this fundamental concept that we cannot but

---

[49] Technically, the descriptive laws of the physical sciences are called "causal" laws – the implication being that they enable us to predict – and therefore "explain" – while the "functional" nature of the biological laws disqualifies them from inclusion in this elevated classification. This makes sense only if one is immersed in both the terminology, and the theory which originated these distinctions, because descriptive laws don't explain things in the usual sense. So the use of alternate terms to denote the concepts has been employed here.

believe in. The solution, it is suggested here, is to accept Hume's critique unequivocally – as, that is, it pertains to the concept of "push" causation – and adopt in its place a full-fledged concept of forward, or "pull" causation, with all of its pre-deterministic properties intact.

Another compelling aspect of Hume's insight into causation, then, is that by rendering untenable the concept of push causation, they make possible the acceptance of an alternate theory of determinism, that describes a non-random universe in which all events tend toward some pre-determined end.

If one adopts a systems' model of scientific explanation, end-state determinism can be considered equivalent to "pull" causation: The universe is arranged in an infinity of dimensions where the lower dimensions are both "contained within," and "held together" by the higher dimensions, and all systems are evolving co-operatively to accomplish the ultimate end of the highest one. All we have to do is figure out what that is.

ABOUT THE AUTHOR

Kathleen Milliere was a student and teacher of analytic philosophy, specializing in the philosophy of science.

APPENDIX A
Astronauts and UFO Disclosure

Reports of UFO sighting by astronauts are not new, but date back to over a decade before the beginnings of space exploration, when astronauts who trained as military pilots report having seen UFOs globally.

For example, Major Gordon Cooper, one of the original astronauts, has gone on record as having seen UFOs in space. He also claims to have seen UFOs on earth ten years before observing them in space, and describes them as metallic discs of varying sizes, which shadowed, flew in fighter formation, and could out-maneuver American planes.

During his final descent after a space mission in 1963, Cooper claims to have been approached by a glowing greenish UFO, just over Perth, Australia. The UFO was picked up on radar, but according to Cooper, when he landed, the press informed him that they would not be permitted to report the story. He describes the situation thus:

> For many years I have lived with a secret, in a secrecy imposed on all specialists in astronautics. I can now reveal that every day in the USA, our radar instruments capture objects of form and composition unknown to us. And there are thousands of witness reports and a quantity of documents to prove this, but nobody wants to make them public. Why? Because authority is afraid that people will think of God knows what kind of horrible invaders. So the password is still: "We have to avoid panic by all means."

For his part, astronaut Neil Armstrong asserts that aliens have a base on the moon, and informed him unequivocally to "Get off and stay off." Buzz Aldrin, who visited the moon in 1969, confirmed the presence of UFOs there, and online searches turn up many more reporters.

Astronaut Edgar Mitchell, who credits a childhood in New Mexico with giving him special insight into UFOs, believes that extra-terrestrials have come to earth to help us avoid wars. The White Sands testing area in New Mexico was apparently teeming with UFO activity, and Mitchell says that military personnel there had divulged the fact that extra-terrestrials could disable nuclear missiles, and had shot them down during testing:

> White Sands was a testing ground for atomic weapons – and that's what the extraterrestrials were interested in... They wanted to know about our military capabilities. My own experience talking to people has made it clear the ETs had been attempting to keep us from going to war and help create peace on Earth.

Mitchell, who made what is reported to be the longest moonwalk (1971) and has a PhD from MIT in astronautics, claims that the government is concealing information about extra-terrestrials. He also believes that disclosure is essential for the survival of our species, given the limited life our sun:

> We are being visited.... It is now time to put away this embargo of truth about the alien presence. I call upon our government to open up, and become a part of this planetary community that is now trying to take our proper role as a space-faring civilization.

APPENDIX B:
Alien Abduction Syndrome

Alien Abduction Syndrome, which is officially classified as a form of Post Traumatic Stress Disorder, is treated by psychologists using routine psychotherapy techniques, and there are two senses in which an individual's memories of a UFO abduction can be said to be "recovered."

The first is not, strictly speaking, a case of recovering anything, but might be better expressed as the "assumption of alien abduction." It resembles the way in which psychologists conclude that when a patient presents (x) number of symptoms typically associated with childhood sexual abuse, but has no memory of any such thing, the patient is nevertheless justified in inferring from the symptoms alone that such abuse has occurred, and the memories simply suppressed – presumably as a defense mechanism of some sort. The trauma of such abuse, in other words, is so severe that the patient's brain subconsciously "forgets" the episode(s).

Psychologists who employ this technique in treating emotionally disturbed patients who seem to have no other cause for their dysfunction test for the probability of suppressed memories by administering a quiz – the idea being that even if the patient does not remember such abuse, the fact that they suffer (x) number of symptoms characteristic of people who have been sexually abused as children means that they can confidently assume that this has happened to them, too, and benefit from the same sort of treatment offered to a patient who does remember such abuse.

In the case of UFO abduction with suppressed memories, you don't have to go to a psychologist, but can take the quiz online. It asks questions like, "Is it routine for people to accuse *you* of being an alien?", and it's a lot of fun (Full disclosure: The author took it and passed with flying colors).

But, questions about accuracy aside, the quiz was not designed to function as a form of entertainment. It was designed for precisely the same purpose that the test for suppressed childhood sexual abuse was designed: To provide psychological help for intensely disturbed individuals who suffer from a defined set of emotional problems which are identical to those suffered by individuals who have been abducted by aliens. In both cases, such individuals are often completely unable to function normally, and desperately require some sort of psychological help if they are to participate in society in even a marginal way.

Although it is not, strictly speaking, correct to say that memories are "recovered" in such cases, the patient does operate on the assumption that an abduction has occurred. The quiz can also, of course, be utilized by UFO cults and fans who are seeking to establish their right to be included in the select group who are "chosen" by extra-terrestrials.

The second sense in which abduction memories can be said to be "recovered," is much more straightforward. It applies to individuals who clearly remember being abducted by aliens, but just can't recall any details of the experience. In cases like these, the abductee generally remembers seeing a UFO, and then suddenly finds himself in a different spot, an interval of time which cannot be accounted for having passed.

Hypnotists have worked with such individuals to "recover" memories of what happened, and the memories are generally believed to be accurate, especially in cases where there are two or more individuals abducted in the same incident, and none of them have any memory of what happened, but all relate the same details under hypnosis.

There is also a class of individuals who, although they are neither psychologically disturbed, nor UFO cult worshippers, and are otherwise sane and rational, have abnormally intense feelings towards UFOs and aliens, and are either obsessed with, or have irrational fears and aversions to the very topic of extra-terrestrials. And while they have no memories of having been abducted, they are, for some inexplicable reason, convinced that something alien-related has happened to them.

Usually, people who fall into this final category will avoid at all costs even *discussing* the possibility that they have been abducted. They also staunchly deny that they are irrational, or "suffering" from anything, insisting that a fear of aliens is completely justifiable, given the terrors associated with ETs. With this category of suppressed-abduction memories, then, the alleged suppressor feels completely normal; it is the people who are acquainted with such individuals who insist that there is something "strange" about them.

APPENDIX C
Extra-Terrestrials and Government Conspiracies

Following Frank E. Stranges ("The Stranger at the Pentagon," 1967), there is an extensive literature in UFO lore discussing the alleged alien from Venus named "Valiant Thor," who is purported to have lived in a concealed apartment at the Pentagon during the Eisenhower administration, and initiated numerous discussions with Stranges, who was a Christian minister in a fundamentalist denomination.

According to the standard account, Valiant Thor landed in the capital with several crew members in 1957 and was taken to the Pentagon by a policeman, where he was secretly visited, via an underground tunnel, by President Eisenhower and Vice President Richard Nixon. Some accounts have him meeting with a variety of the planet's greatest scientists as well.

It is possible to view photographs online of two different individuals both alleged to be Valiant Thor, one of whom can also be seen with his alleged alien assistants. Stranges' books are still in print, and there is copious discussion/summarizing online. The aliens are portrayed as being human in appearance – apart, that is, from their extraordinary attractiveness (according to one account) – but they communicated telepathically, spoke 100 languages, and did not eat or sleep.

Thor – or "Val," as he evidently preferred – is said to have come with a complicated Christian message which assured the minister that his beliefs were both true, and consistent with those of the aliens inhabiting Venus, but seems to have been primarily concerned with imparting information about the potentially catastrophic consequences of developing nuclear technology.

He is alleged to have communicated via "telepathy," and his story is central to ufology because so much of the literature associated with extra-terrestrials is of the same, religious/moral nature. Valiant Thor is, not surprisingly, the subject of a great deal of ridicule even within the UFO community, with the anti-religious contingent making fun of the believers, and the believers insisting that the presence of the Pentagon tunnel "proves" Thor's existence. And although that particular logical leap might be every bit as far-fetched as the idea that Venus is awash with underground alien cities, it doesn't necessarily follow from this that post-war extra-terrestrials did not exist, or that the Eisenhower administration was not mixed up with secret space-related military research in some way or other.

President Eisenhower, after all, inherited the classified, "military" approach to the UFO problem directly from President Harry Truman, and was known to have made cryptic remarks about the danger of the "military industrial complex" – which is generally interpreted as a reference to the covert funneling of billions of dollars from the military budget to a few select private industries, specifically for research into space/ET technology.

It is also widely believed that, as president, Nixon was planning to make public classified military information regarding the existence of UFOs/ETs just before he was forced to resign due to the so-called "Watergate scandal" – which is itself presently undergoing renewed speculation and extensive historical revision. Nixon is also alleged to have shown alien bodies to one of his golfing buddies, the famous TV comedian Jackie Gleason.

According to Gleason's ex-wife, he and Nixon discussed UFOs on a regular basis, and on Gleason's birthday in 1973, the president ditched his security detail and secretly took the actor to a hidden room at Homestead Air Force Base, which contained crashed alien bodies, as a sort of "birthday treat."

Gleason – a UFO fanatic whose extensive library of UFO literature was donated to the University of Miami after his death – did not deny the story. But he did denounce and divorce his chatty wife. And Nixon's Secret Service agent wrote a "confessions" book confirming the president's habit of eluding security, so it is certainly possible that if such a visit had in fact been staged, it might have represented part of a larger, calculated, plan on Nixon's part to establish independent confirmation of the UFO information that he was intending to disclose to the general public. Nixon must have known, at the very least, that such a story was almost certain to be leaked by *someone,* and it is almost impossible to believe that he would run such a risk in the absence of some ulterior motive.

It is also entirely possible that the scandal threatening his impeachment – which appears suspiciously un-scandalous to modern eyes – was simply cooked up in order to discredit Nixon and/or prevent the disclosure.

President Kennedy was also strongly opposed to classifying UFO information, and had been planning to go public about extra-terrestrials shortly before he was assassinated. He was known to have requested an assortment of top-secret, classified information just 10 days before his death, and to have been briefing celebrity journalist Dorothy Kilgallen, whom he had retained to write the story – and who was subsequently found dead in suspicious circumstances, with all of her notes missing.

Kilgallen first discussed UFOs in a 1954 newspaper gossip column, noting, "Flying saucers are regarded as of such vital importance that they will be the subject of a special hush-hush meeting of the world military heads next summer."

Her 1955 story, which was syndicated in various newspapers, including the Washington Post and New York Journal America, went on to disclose an account of a classified UFO crash, and referenced information confirming the recovery of an alien craft and bodies. This information had apparently been provided to her personally at a cocktail party by an undisclosed source generally believed to be Lord Mountbatten, an uncle of England's Prince Charles.

Prince Charles has himself reported seeing a UFO, as have many prominent Americans, including newscaster Walter Cronkite (during his off-record observation of a missile test in the South Pacific, 1950's), Senator Richard Russell (in Russia, 1955), Governor of Arizona Fife Symington (Arizona, 1997), President Jimmy Carter (Georgia, 1969), Congressman Dennis Kucinich (Washington, 1982), and

President Ronald Reagan (two sightings, both in California, 1974).

Kilgallen, a TV celebrity as well as journalist, was preparing a "bombshell" book about President Kennedy's assassination when she was allegedly murdered herself, having previously – and presciently – declared: "That story isn't going to die as long as there is a real reporter alive."

This particular conspiracy is far too involved to address here, but is readily available online. It includes Marilyn Monroe's death as well of that of John Kennedy Jr. – whose own political aspirations were well known, and whose death was presented to the press in a seriously distorted way which not only maliciously implied that he was reckless and not-very-bright, but falsely claimed that he had neither the competence nor the official credentials to fly his plane, and neglected to mention the almost-certain presence of a co-pilot who also mysteriously disappeared at that time. It is also interesting to note that President Kennedy's brother Robert, who served as his Attorney General, had apparently conducted his own secret investigation into the assassination, and concluded that JFK was murdered by a "rogue" element of the CIA.

Robert Kennedy was himself murdered shortly after that, and the controversy surrounding *his* death continues to rage, as his killer, who is still alive, and claims to have no memory of the incident, is frequently alleged to have been brainwashed by the CIA; the details surrounding his shooting are seriously at odds with the official report; several witnesses changed/retracted their stories; and so on and so forth.

At any rate, the rogue CIA faction alleged to be responsible for JFK's death is, predictably, reported by ufologists to be an ET-controlled element (again, names and details all readily available online) and the plot is somewhat complicated, but it involves the covert co-operation of superpowers with extra-terrestrials, and the ever-important imperative of concealing their presence on earth from the general world population.

One of the first alleged murders perpetrated to keep this secret was that of President Truman's Secretary of Defense, James Forrestal, who was committed to a military mental institution following his report of having witnessed a UFO, setting off what seems to have been the first UFO-related "conspiracy."

Truman, who made the original decision to classify all things extra-terrestrial and hand the ET problem over to the military, was also the first president to appoint a Secretary of Defense. And when the newly-appointed Secretary of Defense found himself, while en route to an international political conference, in an airplane shadowed by a UFO, his life was effectively ended.

Accounts of what happened next differ somewhat dramatically, depending upon the teller, but when he was eventually confined to a military mental institution, reported to be suffering from "melancholia," Forrestal declared that he did not expect to get out alive.

Reports of exactly what happened both before and after his confinement vary, but all of them are "troubling." What can be said with certainty is that President Truman fired the eminently qualified Forrestal and replaced him with the eminently unqualified Louis Johnson. Forrestal was relegated to a naval mental hospital, where his visits were "restricted." He was denied access to all of the people he repeatedly requested to see, including his brother and his priest, but visited by Johnson, the man who replaced him as Secretary of Defense; President Harry Truman; and Congressman Lyndon Johnson, who tends to crop up in conspiracy theories quite a lot. He was also permitted to see his wife and sons: Once.

On May 22, 1949, the day before his scheduled release, Forrestal was strangled with his bathrobe belt and pushed out of a 16th story window. His death was officially classified a "suicide," and seems to have set off a wave of reports of UFO-related murders, threats, and conspiracies.

Valiant Thor (t.h.) på besök i High Bridge, New Jersey, 1959. Thor var kontaktperso-
nen Frank Stranges kontakt från Venus. För att vara utomjording är Thor misstänkt lik
en vanlig jordmänniska.

Left: Photo of alleged alien from Venus Valiant Thor (far right) and his assistants; Right: Another photo purported to be the alien Valiant Thor. There are several images online of "movie star" Valiant and his female assistant, which were initially provided by the minister who claims to have befriended him, and in all of them both alleged aliens are not only exceptionally attractive, but also have a distinctly "other-worldly" air about them; they do not appear to be quite human. The "Neanderthal" Valiant image on the right was provided by Phil Schneider, a geological engineer turned whistle-blower who claims to have worked extensively on top secret UFO-related projects, specifically designing underground military bunkers.

APPENDIX D
Hitler and UFOs

Adolf Hitler's obsession with the supernatural, and UFOs and extra-terrestrials in particular, is well documented. He despised Christianity, and identified instead with classical mythology, searching in vain for an ancient race of superior beings whom he believed to be his "true" ancestors. Official history reports his failure to find them, but ufology often cites the "Nordic" race of aliens as not only having found *him*, but having provided Hitler with assistance.

Some accounts have this race of aliens assisting Hitler during the war by providing him with advanced ET technology and helping a secret group of elite Nazi soldiers evacuate a colony of "pure race" Aryans to Antarctica and South America during the final two years of the war, when it had become apparent to everyone, including the fuhrer himself, that Germany had lost. Most accounts also include the claim that Hitler did not commit suicide – recall that his body was never recovered – but escaped to South America with their help.

There is apparently at least some credible evidence to support the claim that Hitler did in fact escape to South (or, as some claim, North) America. And it is at any rate interesting to note that the current renewed interest in the possibility seems to coincide with an increasing awareness of UFOs, as has the escalation of various white supremacy hate groups. There is an historical connection between white supremists, and the belief in UFOs.

Some, if not most, ufologists believe that when aliens created humans – purportedly to use as slaves – they fashioned us in different colors as a sort of "branding," so that they could quickly identify who belonged to whom; there was no hierarchical intent behind making the different races. But although the idea that Nordic aliens created the white, "Aryan" race to be superior is contradicted by both the theory and the empirical evidence, the idea that Hitler *believed* this to be the case is not especially far-fetched.

His white supremist ideology had to be rationalized *somehow* to make it socially acceptable, after all, and Hitler is known to have believed in all manner of occult practices. Moreover, UFOs had been spotted in Germany at that time, and Hitler's engineers were busy designing "flying saucers" of their own, which were apparently seen in skies as far away as New York and London:

Numerous articles containing declassified photos are available online, and according to one account:

The project was called the Schriever-Habermohl scheme.

Rudolf Schriever was an engineer and test pilot, Otto Habermohl an engineer. It was based in Prague between 1941 and 1943. Prisoners of the Allies claimed to have seen this silvery flying saucer, which was about six yards across, on several occasions:"

Caption under declassified photo, above: "Hitler ordered Luftwaffe chief Hermann Goering to develop the super weapon that would change the war...eyewitnesses saw a flying saucer marked with the Iron Cross of the German military flying low over the Thames in 1944."

Caption under photo above: "LOOKALIKE: The Repulsine engine looks suspiciously like a UFO"

APPENDIX E
Eisenhower and the Greada Treaty

The Greada Treaty, which is central to ufology, is said to have consisted in an agreement between President Eisenhower and an alien nation. Although it is supposed to be a top secret document, an internet search will turn up various accounts – one lists it in a straight-forward, completely matter-of-fact manner, under "Treaties," right before the Treaty of Versailles – and others are humorously accompanied by badly photoshopped images of aliens in the Oval Office. The genuine incident, which allegedly took place during the period of time which history records Eisenhower as having "gone missing" from a sudden, unscheduled, 1954 vacation at Palm Springs, involves one of three meetings that Eisenhower was supposed to have attended with alien visitors.

The background story is reported to have been somewhat involved, with several "warring" species of aliens all arriving at the same time and attempting to make deals of one sort or another with Eisenhower – the tall, blond, and attractive Nordic-looking species, in particular, offering to "defeat" the small, homely, and vicious little gray species for President Eisenhower in exchange for American nuclear disarmament.

Eisenhower, it seems, was more interested in acquiring advanced ET technology than in disarming, and eventually made a deal with the diminutive gray group, who claimed that their planet, located in the Orion system, was dying, and they were in desperate need of immediate sanctuary.

Knowing that the American government's greatest fear was of disclosure, the gray aliens are alleged to have agreed to stay hidden on earth in exchange for being permitted to abduct human subjects and mutilate cattle.

According to Michael Salla's 2004 "Eisenhower's 1954 Meeting with Extra-Terrestrials," there are several whistle-blowers on record who were either present at the negotiations which resulted in a treaty, or saw the classified papers – including an engineer employed by one of the companies contracted by the military to build top-secret hide-outs for extra-terrestrials, the psychic who communicated with them telepathically, and a representative of the Vatican.

The Vatican, incidentally, is believed to be preparing for its own disclosure, in which extra-terrestrials will be officially welcomed and invited to join the Catholic Church. But back to Eisenhower's treaty:

> The terms of the 1954 Greada Treaty reached with the Grey ETs as follows: "The Greada Treaty stated that the aliens would not interfere in our affairs and we would not interfere in theirs. We would keep their presence on earth a secret. They would furnish us with advanced technology and would help us in our technological development. They would not make any Greada Treaty with any other Earth nation.
>
> They could abduct humans on a limited and periodic basis for the purpose of medical examination and monitoring of our development; with the stipulation that the humans would not be harmed, would be returned to their point of abduction, would have no memory of the event, and that

the alien nation would furnish Majesty Twelve with a list of all human contacts and abductees on a regularly scheduled basis.

(One) whistleblower source for a Greada Treaty having been signed is Phil Schneider, a former geological engineer that was employed by corporations contracted to build underground bases. Schneider worked extensively on black projects involving ETs. He revealed his own knowledge of the Greada Treaty in the following:

"Back in 1954, under the Eisenhower administration, the federal government decided to circumvent the Constitution of the United States and form a Treaty with alien entities. It was called the 1954 Greada Treaty, which basically made the agreement that the aliens involved could take a few cows and test their implanting techniques on a few human beings, but that they had to give details about the people involved."

APPENDIX F
Declassified Images of UFOs

UFOs were were not always classified as top secret, and the government's presentation of the phenomenon has varied dramatically since they first became generally known to the public. But even when information about them was officially off-limits, there have always been witnesses who are determined to divulge what they know about UFOs and extra-terrestrials.

Retired Major Donald Keyhoe, an American Marine Corps naval aviator, was one of the first pilots to come forward. His book, "The Flying Saucers are Real," first published in 1950 after an article presenting the same material caused a sensation when it came out in "True" magazine, is still in print. According to Keyhoe:

> Since 1949 there has been a steady increase in saucer sightings. Most of them have been authentic reports, which Air Force denials cannot disprove.

His ground-breaking book goes on to chronicle report after report, all of which were disputed by the military. According to Keyhoe, UFO observations first began in about 1945. Initially, stories describing UFOs – then referred to as "flying saucers" – were routinely reported in newspaper articles, and a source of considerable popular interest. There was no attempt to classify or "debunk" any of them until 1949, when the government issued two dramatically different official views of the problem:

> **April, 1949:** "(T)he mere existence of some yet unidentified flying objects necessitates a constant

vigilance on the part of Project 'Saucer' personnel, and on the part of the civilian population.

Answers have been – and will be – drawn from such factors as guided missile research activity, balloons, astronomical phenomena...But there are still question marks.

"Possibilities that the saucers were foreign aircraft have also been considered...But observations based on nuclear power plant research in this country label as 'highly improbable' the existence on earth of engines small enough to have Powered the saucers....The saucers are not jokes. Neither are they cause for alarm.

December, 1949: "The Air Force denied the existence of flying saucers. The official report said:

"It will never be possible to say with certainty that any individual did not see a space ship, an enemy missile, or some other object."

The phrasing of the official statement is confusing, but apparently it means something along the lines of "When you think you see a flying saucer, what you are probably seeing is something else." As to the possibility that aliens might be visiting our planet, the official report provided by Keyhoe is even more confusing, but it seems to imply something like "We have nukes now, so they wouldn't dare":

Such a civilization might observe that on Earth we now have atomic bombs and are fast developing rockets. In view of the past history of mankind, they should be alarmed. We should therefore expect at this time above all to behold such visitations.

The apparent intent behind this approach was for the military to issue a (deliberately?) convoluted, sort-of denial which left open the possibility that the saucers were real, in order to prevent panic, and then re-introduce the evidence gradually, giving the public a chance to become accustomed to the idea before revealing the entire truth, while re-assuring everyone that America had the military might to repel all manner of attack – just in case. In fact, the opposite happened; by that time, the public was not only already accustomed to the idea, but absolutely fascinated, and for the most part, enchanted. And the minute the military issued a denial, people assumed that something awful was being concealed, and became terrified – and even more fascinated.

People tend to define the 50's sensibility towards all things extra-terrestrial by the classic, 5-star movie of that era, "The Day the Earth Stood Still," which depicts a benevolent, persecuted alien who arrives with a message of peace which no-one hears or understands. But, as the movie tragically suggests, the prevailing attitude towards ETs seems to have been squarely in the pro-persecution camp, and for historical accuracy, it's hard to beat the B-rated entertainment of the era.

It is possible to view no-star movies and popular television shows from the 50's in archival sites online, which, in addition to being tremendously entertaining, give a real-time sense of both the terror of, and the fascination with, the subject of flying saucers – with fear of the government often exceeding fear of the aliens.

For example, Googling "Internet Archives" turns up campy titles like "The Flying Saucer: Streaking Out of the Unknown!", and "Invasion of the Saucer Men!", and a common theme was that of the lone individual who is both tormented by the memory of witnessing a UFO event, and pursued by deadly, black-suited undercover agents determined to convince him otherwise – or else.

But note that all of this stands in stark contrast to the initial, strictly scientific approach towards flying saucers. By the mid-1950's, people were beginning to believe the official government line that UFOs were not real after all, and even to ridicule the idea. But at the same time, any serious attempt to address the issue immediately became a political matter, and the result was a move away from fantastic, science fiction-type entertainment, towards thriller-genre films focusing on the perils of attempting to expose "the Truth."

It is impossible to know with any certainty whether or not the general public still believed in flying saucers by the late 1950's or early 60's; it would probably be more accurate to say that most people simply weren't interested in the subject. Note that science has never been a very big draw with the general population, and even before information pertaining to UFOs was classified by the military, inquiries into the mechanics of the matter really only appealed to a select few.

Add to this the fact that virtually no information was available even to them, and it's easy to understand why once the subject lost its novelty, it essentially disappeared from the public radar.

On the following page are reproductions of pre-1950 newspaper and magazine articles which present UFOs in a straight-forward, newsworthy way, without a hint of humor or derision:

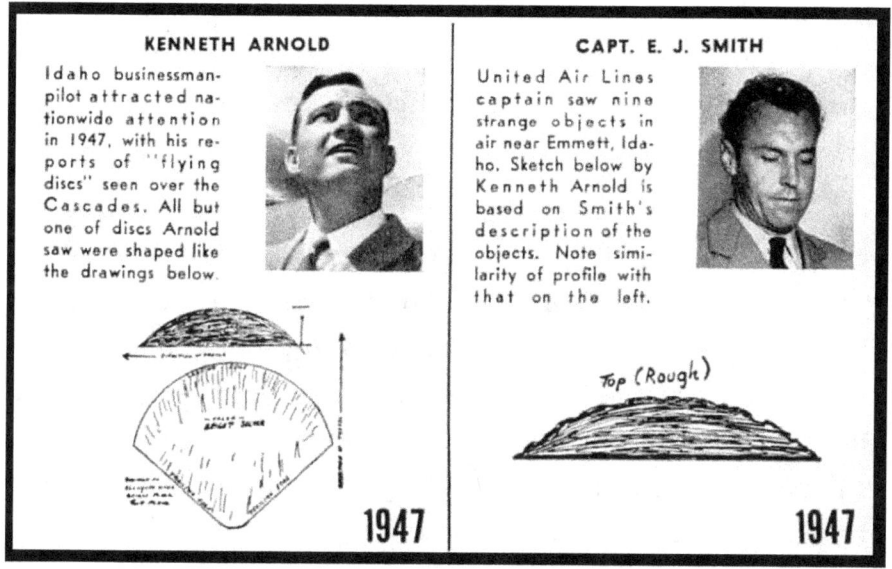

KENNETH ARNOLD

Idaho businessman-pilot attracted nationwide attention in 1947, with his reports of "flying discs" seen over the Cascades. All but one of discs Arnold saw were shaped like the drawings below.

CAPT. E. J. SMITH

United Air Lines captain saw nine strange objects in air near Emmett, Idaho. Sketch below by Kenneth Arnold is based on Smith's description of the objects. Note similarity of profile with that on the left.

Top (Rough)

1947

1947

Serious interest in UFO's, then, has always been confined to a select segment of society; notably those who possess both a keen interest in science, and the lack of other pressing concerns and/or constraints upon their time. In the early days of UFO sightings, it was not unusual for European men in the upper and leisured classes to pursue a serious interest in extra-terrestrials – in much the same way that Victorian gentlemen had, following Darwin's ground-breaking theory of evolution, studiously immersed themselves in the study of geology.

The present queen of England's consort, Prince Philip, for example, who was born in 1921, is said to have a deep interest in UFOs, and this interest is hardly unique among the royals in that country. His son, Prince Charles, is likewise keen. And the late Lord Mountbatten was not only an authority on the subject, but is alleged to have leaked classified British information to an American journalist, and even submitted an official report of a flying saucer landing witnessed on his own estate (reproduced below).

In America, where there is no real leisured class, and the "royals" are entertainers of various ilk, there has not been the same level of interest in UFOs among the wealthy. At least not historically. But this is changing as ufology becomes increasingly focused on the occult, attracting all manner of New-Age Hollywood adherents. And, of course, the "misfit" stigma traditionally attached to UFO witnesses appeals to the artistic, subversive set.

All of this, however, only serves to re-enforce ufology's official status as absurd, and although most celebrity believers in the supernatural are simply dismissed out of hand as "kooks," the many contemporary American actors and musicians who have reported serious UFO sightings are essentially ignored right along with all of the other credible evidence of an extra-terrestrial presence on earth; the topic is no longer current, and in the absence of any threat, space news is typically relegated to the category of boring old science, which has no relevance to either real life, or important human issues. In liberal circles, it is further denigrated as a waste of money and resources, the belief being that America ought to focus on fixing its social problems instead of funding expensive space-related projects and contributing to military excesses.

In contrast, the pro-military administration under Republican president Donald Trump has not only vowed to put man back on the moon, but has secured funding for a fully-public space branch of the military, and even gone on record as having received military briefings on possible UFO sightings. Many ufologists regard this is a precursor to what they expect will be the president's official confirmation of the extra-terrestrial presence on earth, although it is of course questioned whether there will be anything like full disclosure, as it is also widely believed that existing classified military space technology far exceeds anything Americans are prepared to accept or even comprehend.

But although the reality of ET-inspired space weapons and technology like anti-gravity propulsion, teleportation, and telepathic, dimension-busting "super soldiers" is routinely assumed in the UFO literature, the historic relationship between the (secret) space branch of the military and Democratic presidents is reported to be so antagonistic that no Democratic president since Kennedy – who is believed to have been assassinated for refusing to tow the secrecy line – has even been informed about either the extra-terrestrial presence here, or the military technology that it has allegedly provided.

Whether the new Biden administration has been fully informed regarding the present UFO situation is impossible to determine. What is known is that it has inherited an intelligence committee devoted to both discovering, and disclosing classified UFO files:

In June 2020's Intelligence Authorization Act (IAA), the Senate Select Committee on Intelligence (SSCI) authorized appropriations for fiscal year 2021 … and supported … efforts to reveal any links that U(FOs) have to adversarial foreign governments, and the threat they pose to U.S. military assets and installations.

This news was, predictably, buried by the mainstream press, as was the fact that the COVID relief bill, signed by President Trump in August of 2020, contained a clause requiring that the Pentagon release its report to Congress no later than June 25, 2021. There has, also predictably in a party which is adamantly "anti-space," been no mention of UFOs either during the campaign leading up to, or two months into the Biden administration — apart, that is, from the re-affirmation of its commitment to concentrating upon terrestrial and social problems, and assorted jokes about the new Space Force branch of the military. And this despite the fact that the disclosure deadline is less than three months away.

The fact that President Trump was not only fully informed, but actually poised to publicly disclose at least some of what he knows, seems to indicate either the presence of a new threat, or some sort of theoretical breakthrough in understanding the subject; perhaps it is a combination of both. Whatever is happening behind the scenes, however, there is within the UFO community a sense that the problem has finally been addressed intelligently – even masterfully.

The gradual release of de-classified information, for example, has resulted in considerable de-sensitization of the subject, while also serving to re-pique public interest. And when one considers the need for global co-operation in confronting the problem, President Trump's willingness to, and tremendous success in opening communication with other nuclear nations – in particular, North Korea – is both imperative and unprecedented. But note that the contrast between an official disclosure of the problem, when it finally does come – accompanied, presumably, by mind-boggling descriptions of high-tech "solutions" – and historical expressions of wonder at the arrival of flying saucer friends from outer space, could hardly be more pronounced.

For example, below is a reproduction of an official report filed by England's Lord Mountbatten, an early UFO researcher, as it was related to him by his employee, Fred Briggs, in 1955. Briggs claimed to have been knocked off of his bicycle by a strange force on the Mountbatten estate in Hampshire before witnessing the following:

> "The object was shaped like a child's huge humming-top and half way between 20ft or 30ft in diameter," the official report states.
>
> "Its colour was like dull aluminum, rather like a kitchen saucepan. It was shaped like the sketch which I have endeavoured to make, and had portholes all around the middle, rather like a steamer has."
>
> It went on: "While I was watching, a column, about the thickness of a man, descended from the centre of the saucer and I suddenly noticed on it, what appeared to be a man, presumably standing on a small platform on the end.
>
> "He did not appear to be holding on to anything. He seemed to be dressed in a dark suit of overalls and was wearing a close fitting hat or helmet."
>
> Fred added: "As I stood there watching I suddenly saw a curious light come on in one of the portholes.

"It was a bluish light rather like a mercury vapour light.

"Although it was quite bright, it did not appear to be directed straight at me, nor did it dazzle me, but simultaneously with the light coming on I suddenly seemed to be pushed over and I fell down in the snow with my bicycle on top of me.

"What is more, I could not get up as though an unseen force was holding me down.

"The flying saucer then disappeared out of sight "almost instantaneously."

Note the tone of the story, as it was related by the witness. There is no attempt to conceal the incident, and no apparent thought that it would be ridiculed, or disbelieved; it was simply relayed as clearly and thoroughly as possible, the way that one might describe a traffic accident.

De-classified UFO images are cropping up with increasing frequency in the online news, as they are being released in large batches containing various other historical images of unsorted subjects, ranging from movie stars and weapons of mass destruction, to underground UFO bunkers, to the Queen of England unattractively attired in an army uniform, to fun facts about Mount Rushmore, to silly outdated designs for assorted technology, like the really bad seating plan for pilots, reproduced below:

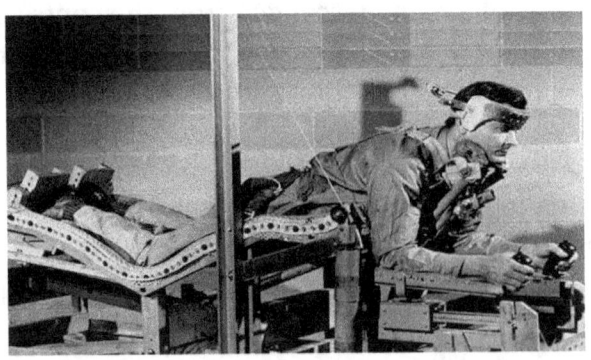

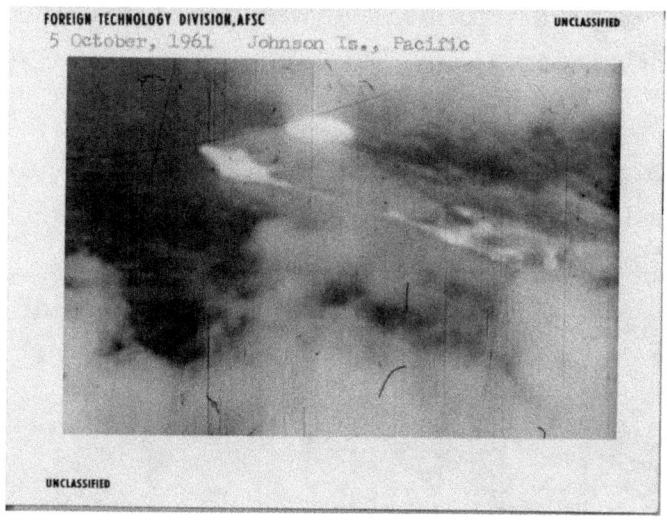

FOREIGN TECHNOLOGY DIVISION, AFSC — UNCLASSIFIED
5 October, 1961   Johnson Is., Pacific
UNCLASSIFIED

The collections also include blurry, unedited, historical images of genuine UFOs (above) alongside pictures of various military aircraft which were designed to *look* like UFOs, with captions indicating that perhaps *this* is what people mistook for a flying saucer (below).

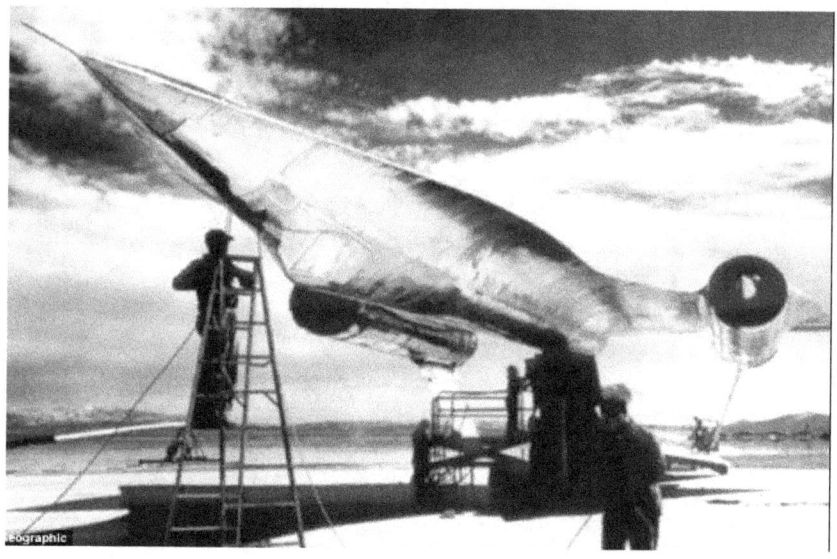

These are in turn followed by images of, say, the first nuclear weapon, Jimmy Stewart, and the Roswell recovery sight.

For example, an article entitled "Declassified Government Images That Will Shock You!", begins with a wacky image of the original underwater diving suit:

It is followed by blurry images of flying saucers and individuals viewing alien crash sites:

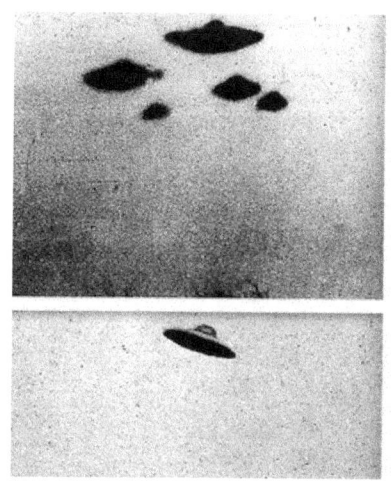 

Next will come something like a photo of a plaster cast model for Mount Rushmore, and various war photos, engineering projects, and the like, followed by more UFO sketches/images:

Finally, there will be images of military projects designed to look like UFO's, complete with captions making the existence of UFOs seem improbable, followed by more non-UFO images and stories, and then straight-forward images and reports of real-live, honest-to-goodness flying saucers. The exercise appears to present UFOs as "historical curiosities," confirming, perhaps, their existence, but making them seem dated and irrelevant.

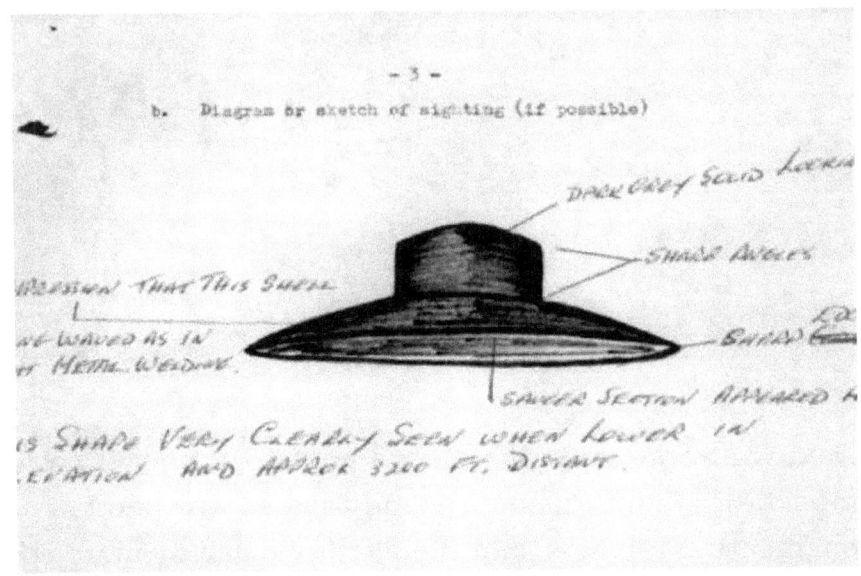

Above, military "flying saucer" circa 1956, followed by the caption: "Meet the Jetsons: It looks like the United States Air Force was very interested in paying the final frontier a visit as newly uncovered documents showed a top-secret program to build America's own flying saucer."

Above: Site of the Roswell crash, followed by a photo of a military-made flying saucer designed to resemble a UFO:

Above: Photo of military-made flying saucer designed to resemble a UFO. Caption below photo: "This is an image of a supersonic flying saucer built by the Airforce. Such an aircraft makes one think that previous sightings may have highly likely been United States government saucers performing secret missions."

Above: Government official inspects a downed flying saucer in a recently declassified document.

Above: Image of flying disc removed from the Roswell crash site. There is a weather balloon attached to the saucer, for unexplained reasons.

As the story goes, in 1947, rancher Mac Brazal found the crash debris, some of which was other-worldly material that could not be cut in two. Army Air Force intelligence agent Major Jesse Martell was enlisted to inspect, and took a sample of the fantastic, indestructible metal home to show his young son. The material was reported to be of pastel shades of pink and purple, and decorated with peculiar symbols of an unknown character.

Regional newspapers from mid-western America carried the story of a crashed flying saucer. After being questioned by the military, Brazal changed his story, and claimed that he had actually recovered a weather balloon, not a flying saucer. Major Martell claimed that there was a cover-up, and newspapers all over America featured his story the next day.

It is impossible to tell, from the image of the crash site, (declassified, above) exactly what sort of vehicle has landed, but the Roswell incident remains one of the most popular UFO tales in history, spawning perhaps thousands of books, documentaries, movies, and one of the longest running television shows ever.

The public, then, is once again acquainted with the topic of UFOs. The difference is that the "fear factor" seems to have disappeared entirely – at least for those who have had no personal contact with extra-terrestrial object or events, and view the subject as a source of entertainment. Abductees, astronauts, and pilots, none of whom find extra-terrestrial events amusing, are starting to speak out, and demand not only answers, but action on the part of the government which is concealing them. The government, for its part, is obliging – at least after a fashion.

On May 27, 2019, the government released an official military photo and story to the New York Times, (reprinted below) which made it appear that full disclosure was imminent.

'Wow, what is that?' Navy pilots report unexplained flying objects'

May 27, 2019 The New York Times"© Adam Ferguson/The New York Times

A U.S. Navy pilot and a weapons system officer from the VFA-11 "Red Rippers" squadron after returning to the aircraft carrier USS Theodore Roosevelt in the Persian Gulf in 2015. The squadron began noticing strange objects just after the Navy upgraded the radar systems on its F/A-18 fighter planes."

WASHINGTON — The strange objects, one of them like a spinning top moving against the wind, appeared almost daily from the summer of 2014 to March 2015, high in the skies over the East coast. Navy pilots reported to their superiors that the objects had no visible engine or infrared exhaust plumes but that they could reach 30,000 feet and hypersonic speeds.

"These things would be out there all day," said Lieutenant Ryan Graves, an F/A-18 Super Hornet pilot who has been with the Navy for 10 years and who reported his sightings to the Pentagon and Congress. "Keeping an aircraft in the air requires a significant amount of energy. With the speeds we observed, 12 hours in the air is 11 hours longer than we'd expect."

In late 2014, a Super Hornet pilot had a near collision with one of the objects, and an official mishap report was filed. Some of the incidents were captured on video, including one taken by a plane's camera in early 2015 that shows an object zooming over the ocean waves as pilots question what they are watching.

"Wow, what is that, man?" one exclaims. "Look at it fly!"

No one in the Defense Department is saying that the objects were extraterrestrial, and experts emphasize that earthly explanations can generally be found for such incidents. Graves and four other Navy pilots, who said in interviews with The New York Times that they saw the objects in 2014 and 2015 in training maneuvers from Virginia to Florida off the aircraft carrier USS Theodore Roosevelt, make no assertions of their provenance. But the objects have gotten the attention of the Navy, which this year sent out new classified guidance for how to report what the military calls unexplained aerial phenomena, or unidentified flying objects.

Joseph Gradisher, a Navy spokesman, said the new guidance was an update of instructions that went out to the fleet in 2015, after the Roosevelt incidents.

"There were a number of different reports," he said. Some cases could have been commercial drones, he said, but in other cases "we don't know who's doing this, we don't have enough data to track this. So the intent of the message to the fleet is to provide updated guidance on reporting procedures for suspected intrusions into our airspace."

The sightings were reported to the Pentagon's shadowy, little-known Advanced Aerospace Threat Identification Program, which analyzed the radar data, video footage, and accounts provided by senior officers from the Roosevelt. Luis Elizondo, a military intelligence official who ran the program until he resigned in 2017, called the sightings "a striking series of incidents."

The program, which began in 2007, was officially shut down in 2012 when the money dried up, according to the Pentagon. But the Navy recently said it investigates military reports of UFOs, and Elizondo and other participants say the program — parts of it remain classified — has continued in other forms. The program has also studied video that shows a whitish oval object described as a giant Tic Tac, about the size of a commercial plane, encountered by two Navy fighter jets off the coast of San Diego in 2004.

Leon Golub, a senior astrophysicist at the Harvard-Smithsonian Center for Astrophysics, said the possibility of an extraterrestrial cause "is so unlikely that it competes with many other low-probability but more mundane explanations." He added that "there are so many other possibilities — bugs in the code for the imaging and display systems, atmospheric effects and reflections, neurological overload from multiple inputs during high-speed flight."

Graves still cannot explain what he saw. In the summer of 2014, he and Lieutenant Danny Accoin, another Super Hornet pilot, were part of a squadron, the VFA-11 "Red Rippers" out of Naval Air Station Oceana, Va., that was training for a deployment to the Persian Gulf.

Graves and Accoin spoke on the record to The Times about the objects. Three other pilots in the squadron also spoke to The Times about the objects but declined to be named.

The pilots began noticing the objects after their 1980s-era radar was upgraded to a more advanced system. As one fighter jet after another got the new radar, pilots began picking up the objects but ignoring what they thought were false radar tracks.

But Graves said the objects persisted, showing up at 30,000 feet, 20,000 feet, even sea level. Then pilots began seeing the objects.

What was strange, the pilots said, was that the video showed objects accelerating to hypersonic speed, making sudden stops and instantaneous turns — something beyond the physical limits of a human crew.

Asked what they thought the objects were, the pilots refused to speculate.

This report was followed by another, on September 18, 2019, officially acknowledging the existence of UFOs:

"UFO videos are footage of real 'unidentified' objects, US Navy acknowledges

For the first time, the U.S. Navy has acknowledged the three UFO videos that were released by former Blink-182 singer Tom DeLonge and published by the New York Times are of real "unidentified" objects.

"The Navy considers the phenomena contained/depicted in those three videos as unidentified," Navy spokesman Joseph Gradisher told The Black Vault, a website dedicated to declassified government documents.

Gardisher added that "the 'Unidentified Aerial Phenomena' terminology is used because it provides the basic descriptor for the sightings/observations of unauthorized/unidentified aircraft/objects that have been observed entering/operating in the airspace of various military-controlled training ranges."

The article went on to express astonishment at the language being employed to officially confirm the "unexplained" nature of the objects in a series of videos taken between 2004 and 2015, which also tended to confirm the official plan to introduce the topic to the public gradually:

> "The first video of the unidentified object was taken on Nov. 14, 2004, and shot by the F-18's gun camera. The second video was taken on Jan. 21, 2015, and shows another aerial vehicle with pilots commenting on how strange it is. The third video was also taken on Jan. 21, 2015, but it is unclear whether the third video was of the same object or a different one.

> John Greenewald, Jr., who publishes The Black Vault, (said) he was surprised at the language the Navy used in its official statement.

> "I very much expected that when the U.S. military addressed the videos, they would coincide with language we see on official documents that have now been released, and they would label them as 'drones' or 'balloons,'" Greenwald told the news outlet. "However, they did not. They went on the record stating the 'phenomena' depicted in those videos, is 'unidentified.' That really made me surprised, intrigued, excited and motivated to push harder for the truth.

The article, which was astonishingly unlike anything ever published previously, given both the tone and content of the official statements contained therein, not only confirmed the existence of UFOs, but was accompanied by a just-declassified reproduction of a military radar image of one of the objects under consideration, along with an acknowledgement that the military had conducted a serious study of the phenomena:

> "The videos in question, known as "FLIR1," "Gimbal" and "GoFast," were originally released to the New York Times and to The Stars Academy of Arts & Science (TTSA). In December 2017, Fox News reported that the

Pentagon had secretly set up a program to investigate UFOs at the request of former Sen. Harry Reid, D-Nev.

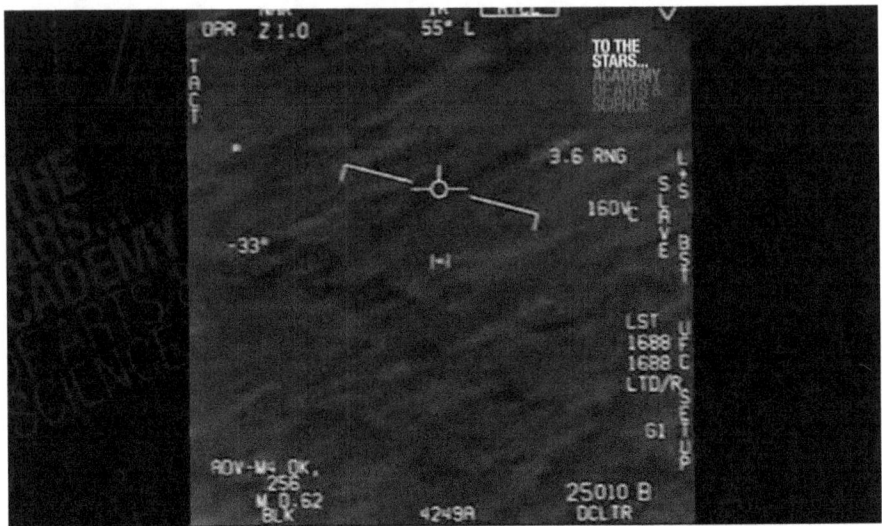

Caption under radar image: "Newly declassified video and audio show U.S. Navy pilots' apparent sighting of alien craft near East Coast."

The article concluded with an appeal to the public to take the objects seriously, and indicated that the Navy had updated its guidelines for making reports, and was stepping up its efforts to keep the government informed on this issue:

> Luis Elizondo, the former head of the Pentagon's Advanced Aerospace Threat Identification Program (AATIP), has previously said that people should pay attention to the comments the government is making about UFOs.
>
> "What the pilots encountered that day was able to perform in ways that defied all logic and our current understanding of aerodynamics," Elizondo wrote...of the 2004 encounter by U.S. Navy pilots who witnessed the object off the coast of San Diego.

Earlier this year, the Navy issued new classified guidelines on how to report such instances "in response to unknown, advanced aircraft flying into or near Navy strike groups or other sensitive military facilities and formations."

The Defense Department also briefed Senate Intelligence Committee Vice Chairman Mark Warner, D-Va., in June, along with two other senators, as part of what appeared to be heightened efforts to inform politicians about naval encounters with unidentified aircraft.

Warner's spokesperson indicated that the senator sought to probe safety concerns surrounding "unexplained interference" naval pilots faced, according to Politico. The outlet reported more briefings were being requested as news surfaced that the Navy revised its procedures for personnel reporting on unusual aircraft sightings."

In this story, it is implied that the then-recently-reported observance of UFOs by military pilots had been made possible by just-developed radar technology. It was also made explicit that there had been a secret branch of the military dealing with extra-terrestrials all along – which was allowed to languish due to lack of funding, but was in 2019 not only fully funded, but fully public.

Note how, although the reporting pilots interviewed here were not given official permission to release information, or even to speculate about the UFOs, far from making the UFO reporters in this story look ridiculous, the opposite is the case, and it is the "debunking" scientist who is made to sound uninformed and almost silly.

The public, then, is beginning to take the subject of UAPs

seriously. In addition, there is also both an historic, and growing, connection between UFOs and religion. Traditionally, the religious angle involved various New Age cults, often led by individuals claiming to "channel" wisdom from various extra-terrestrial sources. The latest focus on the connection between extra-terrestrials and spirituality comes from fundamentalist Christians, who have re-defined various supernatural entities described in the Bible as "extra-terrestrials," and embraced both with enthusiasm.

The internet is over-flowing with podcasters who feature this new, UFO-friendly take on Christianity, and adherents in Congress are demanding UFO disclosure of the military. The movement is premised upon a literal re-interpretation of the Bible, and one notable precursor to this perspective is an extraordinary novel by Norwegian writer Karl Ove Knausgaard, called "A Time for Everything," (Archipelago, 2009) which retells stories from the Bible in psychological/literal terms, presents the reader with a world where what we now know as "angels," were once ordinary inhabitants of our planet, and essentially mirrors what many Christians now believe about the reality of supernatural beings who visited earth in antiquity.

The very latest in this field of research focuses on Catholicism, which was the first Christian denomination. It involves a secular/scientific reading of the Bible, with an emphasis upon exposing "secret knowledge" about all manner of supernatural beings whose existence is allegedly being hidden from the public by the Vatican. Church leaders, it is claimed, falsely "mytholgized" the text, turning what were intended to be metaphysical/scientific teachings into a religion, and "inventing" a divine being who ruled over the masses, punishing the wicked and granting everlasting bliss to the worthy.

The original Bible, which was based upon ancient Sumerian texts, apparently not only espouses re-incarnation, but features a sort of "universal consciousness" which is essentially identical to what is sometimes referred to as an "over soul" in the Eastern tradition, and is accepted by ufologists as a sort of supernatural "epiphenomena" which facilitates telepathic communication and mind-to-objects causation.

But whether it is due to the ultimate rejection of "Boomer" values, or an interest in UFOs, there is no question that Christianity is undergoing a real renaissance among young people, with both highschool and college campuses holding enormous, rave-like revivals featuring "come on down" conversions and full immersion baptisms.

Endorsed by popular internet figures like Elon Musk and Russell Brand, this new take on Christianity is finding empirical confirmation in the present UFO influx, as fascinated believers with cell phone cameras record images of "plasmoid orbs" in the sky which uncannily resemble early images of angels found in the Bible.

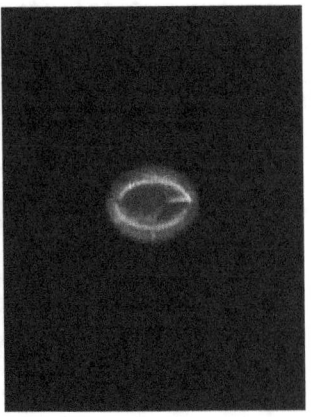

Above, left: Depiction of "Ophanim," or highest order of angels, which appear as wheels in Ezekiel's vision of the "Chariots of the Gods." Right: Internet upload of an "orb" which looks similar to the Biblical angel.

The internet features many uploads of "plasmoid orbs" in the sky, (December, 2024) UAPs which resemble biblical images of angels, and are believed by many Christians to be a precursor to the "second coming" of Jesus promised in the Bible.

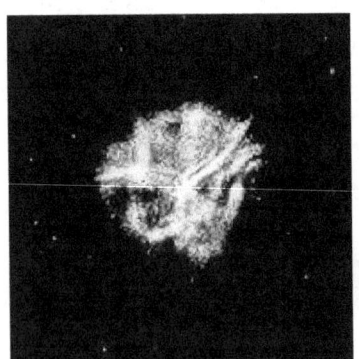

At present writing, (December. 2024) there is a large and growing influx of UAPs hovering over military installations globally, which the government will not, or cannot, explain, leaving the public to come to its own conclusions. Many believe that it is their respective religion's version of the Second Coming. Ufologist Stephen Greer, for his part, insists that it is a false flag alien invasion, being staged by the secret globalist military to terrify the masses into begging their leaders to hand over their sovereignty to an international authority – "capable of defeating the alien menace."

Above: UAPs over San Francisco, Dec., 2024, believed by some to be an alien invasion, and others to represent the Second Coming.

The incursion does not have the slick, contrived, air of a false flag operation. It also lacks the quick and ready "solution" that these tactics are designed to justify. But if it is a fake alien invasion being simulated to control people by terrifying them with "UFOs," then it is failing miserably: Secular UFO advocates online are pleading with the aliens to "Save us from our evil governments!", New Age UFO cult worshippers are dragging their children out into the night to blow kisses of welcome to the "divine visitors," and modern Christian converts are calling up at the high tech lights, "Blink if you know Jesus!"

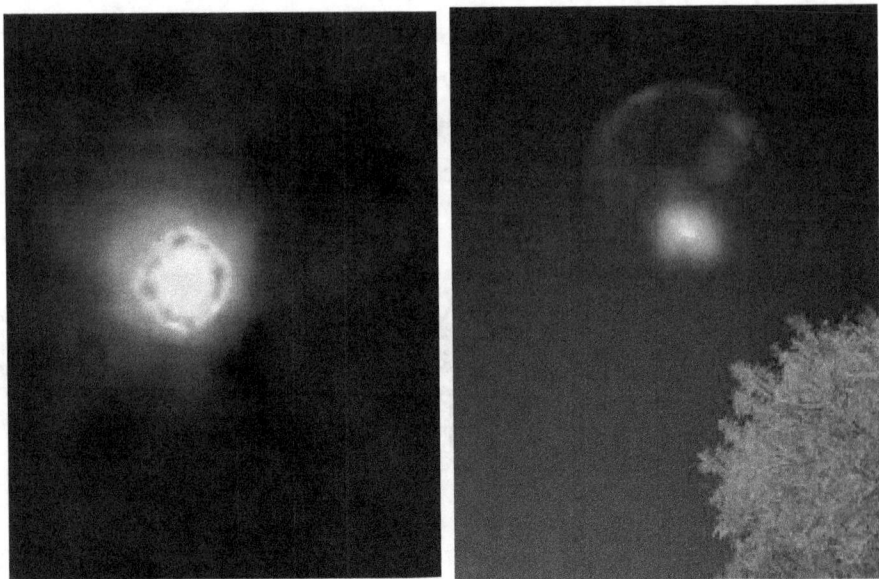

Above: Mysterious, wheel-like "orbs" appear in the sky all over the world, leading many Christians to believe that they are supernatural "angels." heralding the Second Coming.

APPENDIX G
The Darwinian Cultural Revolution

Although it is officially classified as a scientific theory, natural selection appeals to the vast majority of its adherents for purely social/normative reasons. Strident neo-Darwinians outside of academia – which is essentially the only place where the theory is popular – are typically rabidly anti-Christian atheists who would never have latched onto the theory in the first place if it had not been advanced as a means of "defeating" the religion – and especially its moral imperatives, which they despise and denounce as "conservative," and "straight-laced."

Conversely (and predictably) critics of evolutionary theory are almost all religiously devout individuals who also aren't interested in science of any kind, but denounce natural selection *because* it offends their moral sensibilities.

The arguments for and against natural selection, then, essentially reduce to a religious conflict, and if Darwin had advanced the theory only to describe non-human organisms, it is unlikely that any non-specialists would have paid the slightest bit of attention to it, let alone gotten so riled up over the concept. But as it is, let an academically-inclined individual even begin to voice any criticism, and he is liable to be immediately cut off with the sneer, "What are you, some sort of fundamentalist *Christian?*"

This happened to the author on such a regular basis that it was almost routine. Yet once people did listen to dispassionate, purely academic criticism of a logical/scientific sort, they almost without exception expressed astonishment that the theory of natural selection had ever been accepted as scientific in the first place. (Academics who did *not* express astonishment that natural selection was such a bad theory tended to be so bored by the topic that they expressed astonishment that anyone would care one way or the other how bad it was.) It is, it seems clear, primarily non-academics who care about natural selection, and what these people really care about, it also seems clear, is religion, and not science.

It is unquestionably the same conflict between science and religion ushered in by the theory of natural selection which continues to divide the world culturally. In fact, natural selection distinguishes itself, if at all, as perhaps the only scientific theory ever to set off a *cultural* revolution – the effects of which were so profound that they altered fundamental concepts of morality globally, and extended to virtually every single institution, in every society, causing "backlash" outbreaks of fundamentalism which are not only still extant, but flourishing, and perpetuating the original "Darwinian war" between science and religion via terror and intolerance.

This is not the usual consequence of a scientific revolution. The Copernican revolution, for example, didn't do anything much for society at all, despite the fact that the difference between man's being the center of the universe and his being an inconsequential little speck in space could scarcely be more pronounced.

The Darwinian revolution, in contrast, not only secularized forever both art and music, and inspired Dostoyevsky's characters to rant lines like, "If there is no god, then everything in permitted!", producing the universe's greatest works of fiction, but was single-handedly responsible for both the global collapse of religion, and the incommensurable cultural schisms which have characterized virtually every society since, regardless of which side is successfully oppressing the other at any given time, in any particular place. And its effects are still reverberating everywhere, right down to the unbridgeable social divisions between most Democrats and Republicans in America.

It is virtually impossible to overstate the effects of the Darwinian revolution: A single "Intro to Art History" lesson is all that it takes to illustrate the tremendous difference between art the purpose of which is to glorify the Creator, and art which has no purpose beyond its personal creator's unique vision.

Even schools of art, such as the iconoclastic post-Darwinian "Impressionist" movement which now appears so quaint, reflect this radically new, essentially secular, perspective. The wave of literature which originated with the post-Darwinian "superfluous man" is now so common that it defines writing almost as a matter of course. The very concept of "art-for-art's-sake" has its roots in this same scientific revolution, and virtually every principle and institution which characterizes contemporary Western culture can be seen as a direct reaction to the acceptance of Darwin's theory of natural selection.

We tend to think of scientific revolutions as being responsible for purely technological advances, but this one changed humanity – irrevocably and completely –and what the neo-Darwinian is really passionate about has nothing whatever to do with science; he is defending his God-given secular right to "cultural superiority."   So presumably it will require an even bigger revolution to dislodge him, and a post-terrestrial perspective might be just what's wanting.

Unfortunately, the same virulently anti-religious sentiments which have quashed criticism of evolutionary theory for so long also come into play when evidence of extra-terrestrial events is advanced, as so much of ufology is infused (or, according to the devoutly secular set, "infested") with not only Christianity, but New-Age beliefs of various sorts, spiritualism, and the supernatural.

It is exceedingly difficult for an anti-religious individual to even read the literature, let alone listen to UFO cult members espousing spirituality as an alternative to science. And scientists, who aren't trained in logic, not only find the entire business intolerable, but lack the tools to analyze *any* UFO claim effectively. The result is that even pro-UFO specialists like J. Allen Hynek discredit on principle the vast majority of the evidence, simply because they can't stand the way that it sounds.

But logically speaking, if all of the available evidence consists in second-hand reports, there are no grounds for believing the fellow who claims to have met a clandestinely-trained super soldier who teleported through a Stargate tunnel into a future dimension where he battled hostile aliens using their own reverse-engineered ET advanced-weapon technology, while rejecting the claim of the blissed-out religious cult member who insists that aliens imbued him with nirvana and lowered his blood pressure.

What we desperately need are some new metaphysical principles which will help us make sense of *all* of this – and with any luck also eliminate the false science/religion dichotomy which was Charles Darwin's *true* legacy. These subjects were designed to fulfill different functions, and cannot logically be pitted against each other.

APPENDIX H

The Ontology of Minds and a Brief History of Research in the Field of Telepathy

Research into the use of telepathy, or mind-to-body causation, has a long history, not all of it successful. The projects dealing with strictly physical aspects of brain behavior have met with remarkable success, while those dealing with "mind," or immaterial thought processes, have not. Much of the research in this field was initiated by army scientists who were interested in exploring the possibility that humans might put telepathic ability to use in military operations, and there is considerable literature dedicated to the subject – which is not confined to speculation about UFOs, but is generally purported to have been undertaken by researchers internationally, post-World War 2.

According to some sources, the CIA began research into telepathy in 1952, with a project called "Moonstruck," which involved abducting human subjects and implanting electronic devices into their brains and teeth. (This is the sort of alleged experiment to which the pro-UFO camp alludes when it insists that most abductions are carried out not by ETs, but by the military.)

The experiments then escalated to include the use of electroshock and various mind-altering drugs, while the Soviet Union did things like administer microwaves to its subjects – all for the purpose of discovering a method for controlling thoughts remotely.

They also experimented on lobotomized monkeys,

incorporating the use of things like radio waves and electromagnetic fields, and even attempted to control people's dreams via a technique called "dream telepathy." And neither has the research ceased. Apparently, it is ongoing, and immensely more sophisticated, as medical researchers use computer technology to identify the brain signals associated with muscle movement.

From a philosophical point of view, in contrast, a thought is an inherently personal, or "private," non-physical event, while the neural firing *associated* with thought is a "publicly-observable" and completely un-mysterious physical process, which one would expect scientists to be able to manipulate with advanced technology. Note that the neural identification of particular types of muscle movement was made by long, laborious observation of trial-and-error brain stimulation. The researchers involved knew exactly what sort of material process they were looking for, and how it could be exploited when they found it; they just had to figure out a successful way to do it.

All of this experimentation evidently originated with the belief in telepathy; the only difference is that the original experiments were designed to develop/exploit "natural" telepathy, and the more modern versions concentrate upon what is termed "synthetic telepathy." To the philosopher, this distinction is so significant as to render it an ontological one. Note the tremendous metaphysical difference between the (failed) attempts to locate and harness the power of an hypothetical immaterial telepathy-enabling substance which occurs naturally in brains, and the technologically-successful identification and exploitation of purely physical neural processes.

According to the ontological model proposed here, humans are collectively-controlled individual components of the greater thermodynamic system/entity which contains them, and our apparent ability to control bodily actions telepathically by will is an illusion; the correspondence between what we think of as autonomy, or "wilful thought," and bodily action, represents rather an "awareness" of our part in the pre-determined plan.

This awareness gives the individual the ability to know in advance what he is pre-determined by the system's end-state to do, making it seem as though he is in control of his bodily actions. But note that this sense of feeling in control of one's bodily actions is not innate; it is completely contingent upon social conditioning and prior theoretical beliefs. Moreover, not everyone experiences a sense of autonomy, and those who do, experience it only to various degrees, even when they have been taught to believe in free will – or human agency, as it is sometimes called – and have grown up in societies which presuppose it in all of their family and social relationships, cultural practices, and legal institutions.

Despite its status as the veritable cornerstone of humanity, most people do not believe absolutely in free will, and many do not believe in it at all. Historically, of course, people only felt free when personal accomplishment was involved, blaming evil spirits for any outstanding misdeeds, and mindlessly carrying out as the bulk of their behavior whatever was expected of them. Substitute "negative social influences" for evil spirits, and the situation today isn't much different – except, of course, that many people still believe in evil spirits, and forgiveness has gone completely out of fashion as we increasingly attribute bad behavior to an hierarchy of incurable conditions ranging from mild spectrum disorders to full-blown psychopathy. Nobody, it seems, actually believes in human freedom; people really only believe that their thoughts are causing their bodies to do their bidding – "telepathically," to put it technically.

In the model proposed here, the ability to communicate telepathically with *others,* and not just with with one's own body, would simply represent an enhanced awareness – one which extends to being able to predict what other individuals in the system are pre-programmed to do. "Consciousness," in this model, is the ability to predict what we, personally, have been pre-programmed to do in co-operation with other individuals in the system. From our vantage point as humans, this reduces – or, more accurately, it appears to *us* – as an awareness of the relationship between our own thoughts and bodily actions.

To a telepathic human or an ET, consciousness is expanded to include the awareness of *others'* thoughts as well as one's own, and the corresponding ability to predict their actions in the same way. To a lower-dimension entity like a cell, in contrast, which lacks the structural complexity which makes consciousness possible, there is no awareness at all; there is only pre-determined, co-operative interaction between individual cells, which is done solely in order to fulfill the function of the system which contains them.

To a telepathic ET, the limited awareness of humans might make us seem "unconsciousness" in much the same way that cells seem, to humans, unconscious. But it does not follow from this that the behavior of non-conscious entities is pre-determined, while the behavior of conscious entities is somehow "free." We are all components in systems, behaving co-operatively in order to fulfill the function of the system which contains us, and there is no logical reason to suppose that the laws of physics operate differently upon components with complex structures and/or consciousness.

APPENDIX I
Freedom, Determinism, and Consciousness

The concept of human freedom, or "agency," is closely connected with the concept of "consciousness," as it is generally believed both that humans have free will, and that free will is dependent in some way upon being conscious. In other words, to freely choose its course of action, an organism must, after some fashion, be aware of both the world around him, and his place in it.

In order to explain the obviously non-random interaction of entities like cells, it has been suggested that non-conscious entities might also have "awareness" – not equivalent to that of humans, but relative to their specific circumstances. In the model proposed here, even higher-order, human consciousness does not come with intentionality, agency, free will, or any of the other "push causation" concepts traditionally associated with human behavior.

As theories, these concepts "work" even though they cannot withstand logical scrutiny, so they can still be used meaningfully in everyday discourse, much in the same way that we speak of the sun as "setting." But from a theoretical point of view, end-state determinism, or teleology, entails that *all* behavior, in every dimension, tends toward collectively maintaining the thermodynamic function of the system in which the individual "behaver" is contained.

Human consciousness, in this model, is simply a function of sufficient complexity; it does not come with free will because free will – or "individualism" of any sort – is unintelligible in a teleological system. In fact, aliens from higher dimensions might very well classify us as "*unconscious*" because we are too limited to use the immaterial substance to communicate telepathically with anything but our own bodies.

We are not, then, telepathically connected with the rest of the system which contains us. We are also insufficiently intelligent to have any awareness of the end-state that our individual actions are pre-determined to accomplish. We do have, however, sufficient awareness of our *place* in the system to know (Read: "Predict") what our own bodies will do. Knowing this in advance makes it appear as though we are somehow "choosing" our behavior, and this in turn leads to the idea that humans are "free," while animal behavior and physical events are all determined. But the latter is, however instinctive, logically untenable.

To the (analytic) philosopher, the concept of human freedom has always been problematic, quite apart from the ontological divide between mind and body. Why should human events be uniquely undetermined in an otherwise deterministic universe? Adopting a metaphysical model which incorporates the concept of "pull causation," or pre-determinism, it is suggested, can solve this problem.

In a pre-deterministic, or teleological, universe, the system's ultimate end-state pre-determines every action every individual in that system will take, so it is possible, at least in principle, to know in advance what any particular action will be. "Choice," in this model, simply indicates an awareness of what the individual has been pre-determined to do at a given time. Presumably, higher-dimension entities with greater complexity will also have an enhanced awareness of the systems in which they are contained, and be able to predict with similar accuracy what other individuals in the system will do.

APPENDIX J

Sample UFO Reports

Information pertaining to select incidents of UFO reports was released to the public when the government's "Blue Book" was de-classified, and it is possible to read lists of similar incidents online.

To give an idea of what the available evidence for UFOs looks like, reproduced below are three separate accounts of the famous Ellsworth incident, which took place at Ellsworth Air Force Base near Rapid City, South Dakota, in 1953: The original, verbatim, report; a contemporary pro-UFO "conspiracy" account which concludes with a (contextually distorted) summary by UFO investigator J. Allen Hynek; and the official, "Blue Book" government report. The pilot's report would be considered "evidence," the pro-UFO report, a close, if overly enthusiastic, interpretation of the incident, and the government's report, intentionally inaccurate "narrative":

Ellsworth Airforce Base UFO Incident, 1953

Pilot's Report

I first heard about the sighting about two o'clock on the morning of August 11,1953, when Max Futch called me from ATIC. A few minutes before, a wire had come in carrying a priority just under that reserved for flashing the word the U.S. has been attacked. Max had been called over to ATIC by the OD to see the report, and he thought that I should see it. I was a little hesitant to get dressed and go out to the base, so I asked Max what he thought about the report. His classic answer will go down in UFO history, "Captain," Max said in his slow, pure Louisiana drawl, "you know that for a year I've read every flying saucer report that's come in and that I never really believed in the things." Then he hesitated and added, so fast that I could hardly understand hini, "But you should read this wire." The speed with which he uttered this last statement was in itself enough to convince me. When Max talked fast, something was important.

A half hour later I was at ATIC – just in time to get a call from the Pentagon. Someone else had gotten out of bed to read his copy of the wire.

I used the emergency orders that I always kept in my desk and caught the first airliner out of Dayton to Rapid City, South Dakota. I didn't call the 4602nd because I wanted to investigate this one personally. I talked to everyone involved in the incident and pieced together an amazing story.

Shortly after dark on the night of twelfth, the Air Defense Command radar station at Ellsworth AFB, just east of Rapid City, had received a call from the local Ground Observer Corps filter center. A lady spotter at Black Hawk, about 10 miles west of Ellsworth, had reported an extremely bright light low on the horizon, off to the northeast. The radar had been scanning an area to the west, working a jet fighter in some practice patrols, but when they got the report they moved the sector scan to the northeast quadrant There was a target exactly where the lady repored the light to be. The warrant officer who was the duty controller for the night, told me that he'd studied the target for several minutes. He knew how weather could affect radar but this target was well defined, solid, and bnght." It seemed to be moving, but very slowly. He called for an altitude reading, and the man on the height-finding radar checked his scope. He also had the target – it was at 16.000 feet.

The warrant officer picked up the phone and asked the filter center to connect him with the spotter. They did, and ihe two people compared notes on the UFO's position for several minutes. But right in the middle of a sentence the lady suddenly stopped and excitedly said, "It's starting to move – it's moving southwest toward Rapid."

The controller looked down at his scope and the target was beginning to pick up speed and move southwest. He yelled at two of his men to run outside and take a look. In a second or two one of them shouted back that they could both see a large bluish-white light moving toward Rapid City. The controller looked down at his scope, the target was moving toward Rapid City. As all three parties watched the light and kept up a steady cross conversation of the description, the UFO swiftly made a wide sweep around Rapid City and returned to its original position in the sky.

A master sergeant who had seen and heard the happenings told me that in all his years of duty – combat radar operations in both Europe and Korea – he'd never been so completely awed by anything. When the warrant officer had yelled down at him and asked him what he thought they should do, he'd just stood there. "After all," he told me, "what in hell couldf we do – they're bigger than all of us."

But the warrant officer did do something. He called to the F-84 pilot he had on combat air patrol west of the base and told him to get ready for an intercept. He brought the pilot around south of the base and gave him a course correction thai would take him riglit into the light. which was still at 16.000 feet. By this time the pilot had it spotted. He made the turn, and when he closed to within about 3 miles of the target, it began to move. The controller saw it begin to move, the spotter saw it begin to move and the pilot saw it begin to move – all at the same time There was now no doubt that all of them were watching the same object.

Once it began to move, the UFO picked up speed fast and started to climb, heading north, but the F-84 was right on its tail. The pilot would notice that the light was getting brighter, and he'd call the controller to tell him about it. But the controller's answer would always be the same, "Roger, we can see it on the scope."

There was always a limit as to how near the jet could get, however. The controller told me that it was just as if the UFO had some kind of an automatic warning radar linked to its power supply. When something got too close to it, it would automatically pick up speed and pull away. The separation distance always remained about 3 miles.

The chase continued on north out of sight of the lights of Rapid Cty and the base – into some very black night.

When the UFO and the F-84 got about 120 miles to the north, the pilot checked his fuel; he had to come back. And when I talked to him, be said he was damn glad that he was running out of fuel because being out over some mighty desolate country alone with a UFO can cause some worry.

Both the UFO and the F-84 had gone off the scope, but in a few minutes the jet was back on, heading for home. Then 10 or 15 miles behind it was the UFO target also coming back.

While the UFO and the F-84 were returning to the base – the F-84 was planning to land – the controller received a call from the jet interceptor squadron on the base. The alert pilots at the squadron had heard the conversations on their radio and didn't believe it. "Who's nuts up there?" was the comment that passed over the wire from the pilots to the radar people. There was an F-84 on the line ready to scramble, the man on the phone said, and one of the pilots, a World War II and Korean veteran, wanted to go up and see a flying saucer. The controller said, "OK, go."

In a minute or two the F-84 was airborne and the controller was working him toward the light. The pilot saw it right away and closed in. Again the light began to clirnb out, this time more toward the northeast. The pilot also began to climb, and before long the light, which at first had been about 30 degrees above his horizontal line of sight, was now below him. He nosed the '84 down to pick up speed, but it was the same old story – as soon as he'd get within 3 miles of the UFO, it would put on a burst of speed and stay out ahead.

Even though the pilot could see the light and hear the ground controller telling him that he was above it, and alternately gaining on it or dropping back, he still couldn't believe it – there must be a simple explanation He turned off all of his lights – it wasn't a reflection from any of the airplane's lights because there it was. *A reflection from a ground light, maybe.* He rolled the airp!ane – the position of the light didn't change. *A star* – he picked out three bright stars near the light and watched carefully. The UFO moved in relation to the three stars. Well, he thought to himself, if it's a real object out there, my radar should pick it up too; so he flipped on his radar-ranging gunsight. In a few seconds the red light on his sight blinked on – something real and solid was in front of him. Then he was scared. When I talked to him, he readily admitted that he'd been scared. He'd met MD 109's, FW 190's and ME 262's over Germany and he'd met MIG-15's over Korea but the large, bright, bluish-white light had scared – he asked the controller if he could break off the intercept.

This time the light didn't corne back.

What he UFO went off the scope it was headed toward Fargo, North Dakota, so the controller called the Fargo filter center. "Had they had any reports of unidentified lights?" he asked. They hadn't.

But in a few minutes a call came back. Spotter posts on a southwest- northeast line a few miles west of Fargo had reported a fast-nioving, bright bluish-white light.

This was an unknown – the best..

The sighting was thorougly investigated, and I could devote pages of detail on how we looked into every facet of the incident; but it will suffice to say that in every facet we looked into we sawnothing.

Nothing but a big question mark asking what was it."

Captain Edward J. Ruppelt

The Government's "Blue Book" Report

The case file in USAF's Project Blue Book archive:

This is the official USAF investigation report on the case, Project Blue Book Status Report 12, Page 20 to 23.5 August 1953

Rapid City, South Dakota

Description

Since this sighting was a combined air-visual, ground-visual, air radar, and ground-radar report, it was decided that Project Blue Book would send an investigator to the scene. The controller on duty at the time or the incident was interviewed. His account of the incident was almost identical to that given in the initial TWX. He was on duty at 2005 MST when a GOC post observer called in an unidentified flying object sighted northeast of her post at Blackhawk, South Dakota. (Note: Sunset 1920 MST Twilight 33 minutes.) She reported through the Rapid City Filter Center. She reported that the object was stationary, then moved south toward Rapid City. When the controller got the report that the object or light was headed toward Rapid City, he sent 3 airmen from the radar site to look for it visually. They reported a light moving from generally north to south at a high rate of speed. At this time the controller observed 2 blips going south on the scope. He could not get a distinct track because or ground clutter in the area. In a few minutes the GOC post in Blackhawk called in that the light was back in nearly its original position. An airborne F-84 was vectored into the area and after a search made visual contact. The F-84 was vectored into the blip that was remaining stationary at about 15 miles northeast of Blackhawk. As soon as the F-84 landed, another F-84 took off for CAP. Just about that time, the Blackhawk GOC post called the third time stating that the object was back again. Nothing was on the scope (there was possibly a target in the ground clutter), so the F-84 was vectored in on the visual report. The pilot soon got a visual and started an intercept About that time, the controller picked up both an unknown target and the F-84. Both were on a heading of about 360 degrees magnetic. The blip seemed to stay about 5-10 miles ahead of the F-84. The chase continued until the aircraft was about 80 miles out, then the intercept was broken off. The target continued off the scope. At this time the Bismarck Filter Center was alerted to look for unidentified flying objects. When the pilot got back over the base, he saw another light. This was not picked up on the scope, but the controller did get a return on the height tinder equipment in the general direction of the light, it was 8000 feet. At 0023 MST, Bismarck began to call in reports.

The pilot who was on the first CAP was interviewed next. He stated that he had been making passes at a B-36 north of Rapid City when GCI called and said they had a target west of Rapid City. He searched for about 20 minutes west and south of Rapid City but saw nothing. He returned to base and was about to land when he observed a light northwest of the base. He started out on a heading of 350 degrees magnetic, the object was high (30 deg -45 deg) at 11 o'clock from him. He checked the possibility of a reflection and determined that this was not the cause. He continued his course keeping the object at 11 o'clock for a better view. After about 30 seconds, it disappeared then reappeared for another 30 seconds at the end of which it abruptly faded and was not seen again. The object was silver in color and varied in intensity. It appeared to "pull away" because it got smaller. The comment as to size was that it was "brighter than the brightest star I've ever seen."

The pilot who flew the second CAP was interviewed next. He stated that he took off and started to climb when GCI told him that GOC had a light. He was north of Ellsworth AFB on a heading of 360 degrees magnetic when he saw a light 30-40 degrees to his right and level.

He thought it was a star or planet but as he looked away it appeared to "jump" 15-30 degrees in elevation. (Note: Due to the speed of the aircraft and the fact that the pilot wan intent on identifying the object, he was not exactly sure of his positions. All positions are subject to some error.) The light seemed to be paralleling his course. The first thing the pilot did was to check for reflections in the cockpit (i.e., canopy, gunsight head, etc.). He was sure the light was no reflection in the aircraft. The light, which the pilot estimated to be considerably brighter than a star, changed intensity and changed in color from white to green. When the object was first sighted, the aircraft was at 15,000 feet. The pilot started to climb and the light appeared to climb faster. This was because the angle of elevation increased. He climbed to 26,000 feet. All this time both the radar blip of both the object and the aircraft were being carried and the pilot was talking to the controller on UHF. As the pilot turned into the light on his initial sighting, he turned on his radar gunsight. As he swung onto the target, the warning light came on. No range was obtained since the sight starts to measure at about 4,000 yards. All this might indicate was that something was beyond 4,000 yards. The light remained on until the chase was broken off. After the chase, on the way home, the light blinked on and off several times indicating a possible malfunction. The sight was not checked by maintenance on return and had not been checked since.

The F-84 chased the light for about five minutes, or to about 80 miles north of the base. The light appeared to make slow changes in color and intensity. The pilot stated that the light definitely moved in relation to the stars. After the intercept was broken off, the aircraft returned toward base.

About 20 miles out of base he got a visual on a similar light that changed from red to white. He was on a heading of 180 degrees magnetic at 12-14,000 feet and the light was 10 degrees low to the right. He thought it was a car going around curves in the hills but changed his mind when the red and white lights were of equal intensity. This target was in the ground clutter of the radar but something at 8000 feet was picked up on the height finder radar. The light slowly went out then came back in. It seemed to be west moving since the aircraft was kept on a constant heading and the angle or azimuth and elevation increased. The light was first observed for 30 seconds, it faded, reappeared, then faded again after 30 seconds.

As the pilot came around the west side of the air base and up the east side, he saw another light and turned into it to take gun camera photos. (The photos were no good).

Discussion

A visit was made to the Weather Bureau station at the Rapid City Municipal Airport to check weather and balloon launches (Note: The air base launches no balloons). The observer on duty looked up the balloon track for the balloon launched at 2000 1ST on 5 August 1953 and it went south from the Municipal Airport. This puts it out of the area of the sighting. Data on inversions was not available as it had been forwarded to Asheville, North Carolina. (Note: The balloon tracks and weather for 2000 MST on 5 August has been requested from Asheville.)

No attempt was made to contact the GOC observers at Blackhawk. They had been interrogated by base personnel and were "all excited". It was believed that an investigator talking to them would only further excite them needlessly. All the sightings at Bismarck are doubtful. The AC&W Station called the Bismarck Filter Center and told them to "look for flying saucers", a perfect set up to see every star move around.

The upper air research balloon tracks at Lowry were checked. Two balloons were lost and could have been in the area at the time or the sighting.

A few comments on the sources can be made:

Controller left the impression that he was trying to prove the existence of an unidentified flying object. It is very unfortunate that no scope photos were available to collaborate his story. He saw targets on the scope, there is no doubt about it, but whether they acted exactly as the stated is unknown.

The two airmen that went outside to observe the object that was being carried on radar and reported by the GOC were not sure of what they saw, at least this is the impression they left. They were told to go out and look for a light so they saw one. Their description fits that of a meteor. They only saw a "streak" in the sky. They did not see it return north, only go south.

The first pilot only got a glimpse of a light, so he could not add much.

The second pilot gave the impression of being "on the ball". He obviously was trying to convince himself the light was a star, but was having difficulty. He took a realistic approach and had done some logical reasoning. He was worried about the fact that the light moved relative to the stars.

By eliminating doubtful sightings, the only thing that can be reasonably assured is that a GOC post observed a light. This could be a balloon or star. Radar picked up something in the general area of the GOC post and vectored an aircraft toward it. The pilot saw a light and chased it. He got a radar lock on it, but this could have been a mal-function. The star Capella is possibly visible low on the horizon to the north and the pilot could have seen this. Pending further study, this incident is carried as Unsolved.

Conclusion

Unsolved. UNCLASSIFIED

Pro-UFO, "Conspiracy" Interpretation

Pro-UFO Conspiracy Summary and Analysis:

On August 5 and August 6, 1953 the US Military investigated a UFO incident in Bismarck, North Dakota.

1953 Ellsworth UFO sighting ,Important Radar-Visual UFO

posted on Oct, 31 2009 @ 05:17 PM

What has become known as the Ellsworth Case is one of the most significant radar-visual cases in the annals of UFO sightings.

A lady spotter at Black Hawk, about 10 miles west of Ellsworth, had reported an extremely bright light low on the horizon, radar had been scanning an area to the west, tracking a jet fighter in some practice patrols, but when they got the report they moved the sector scan to the northeast quadrant, there was a target exactly where the lady reported the light to be. The warrant officer who was the duty controller that night studied the target for several minutes, it was at 16.000 feet. The warrant officer called to the F-84 pilot he had on combat air patrol west of the base and told him to get ready for an intercept. By this time the pilot had it spotted. He made the turn, and when he closed to within about 3 miles of the target, it began to move away, the UFO picked up speed fast and started to climb, heading north, the F-84 was on its tail.

The chase continued on north, when the UFO and the F-84 got about 120 miles to the north, the pilot checked his fuel; he had to come back.

Both the UFO and the F-84 had gone off the scope, but in a few minutes the jet was back on, then 10 or 15 miles behind it was the UFO target also coming back.

A second plane was launched and the pilot saw it right away and closed in, he switched on his radar-ranging gunsight. In a few seconds the red light on his sight blinked on - something real and solid was in front of him, he asked the controller if he could break off the intercept.

When the UFO went off the scope it was headed toward Fargo, North Dakota.

Bellow is a report by J. Allen Hynek

In 1953, the year of the Robertson report, there occurred one of the most puzzling cases that I have studied.

At approximately the same time, unidentified blips showed up on the radarscope at Ellsworth Air Force Base, which is near Black Hawk. An airborne F-84 fighter was vectored into the area and reported seeing the UFO's. The pilot radioed that one of the objects appeared to be over Piedmont S. Dak., and was moving twice as fast as his jet fighter. It was "brighter than the brightest star" he had ever seen. When the pilot gave chase, the light "just disappeared." Five civilians on the ground, who had watched the jet chase the light, confirmed the pilot's report.

As the object sped off to the north, Ellsworth Air Force Base notified the spotter's control center in Bismarck, 220 miles to the north, where a sergeant then went out on the roof and saw a UFO. The Air Force had no planes in Bismarck that could be sent after the UFO, which finally disappeared later that night.

The above "conspiracy" account, which is clearly calculated to demonstrate a cover-up of unquestionable UFO activity, conveniently omits Hynek's conclusion:

> I investigated this reported sighting myself and was unable to find a satisfactory explanation. In my report, I noted that "the entire incident, in my opinion, has too much of an Alice in Wonderland flavor for comfort.

Edward Ruppelt, the pilot who reported the incident, was so fascinated by what he had seen, and so frustrated by the official handling of the report, that he not only published a feature in "True" magazine lampooning the procedure, but wrote books about the subject which are generally held in such high regard that he is frequently referred to as a "legend." (See especially his well-received "Report on Unidentified Flying Objects," first published in 1956 and still available both online and through booksellers.) His official reaction to the event, in his own words, is as follows:

> What was the object? For two years, from 1951 to 1953, I flew 200,000 miles, conferred with dozens of top American scientists and an exotic collection of hot-eyed screwballs, stumbled through Florida mangrove swamps, dragged myself out of bed at 3 a.m. to answer transatlantic telephone calls, inspected scores of strange photographs and watched one short amateur movie ninety-seven times in an effort to answer this and similar questions.

Captain Ruppelt, who apparently died of a heart attack at the age of 37, would go on to become not only an expert on the subject of UFOs – it was he who coined that term, finding it both more accurate and a lot less silly than "flying saucer" – but head of what was to become the government's "Project Blue Book," which was originally designed as a serious investigation into the problem of unidentified flying objects, but eventually morphed into a much-maligned and almost comical effort to suppress and debunk all UFO-related evidence, concluding finally, in its "Condon Report" (1968):

1. No UFO reported, investigated, and evaluated by the Air Force was ever an indication of threat to our national security;

2. There was no evidence submitted to or discovered by the Air Force that sightings categorized as "unidentified" represented technological developments or principles beyond the range of modern scientific knowledge; and

3. There was no evidence indicating that sightings categorized as "unidentified" were extraterrestrial vehicles.

Ruppelt, at least initially, was scathingly critical of the way in which the subject was handled. The title of his article in True Magazine (1957) says it all:

"Why Don't The Damn Things Swim so we can turn them over to the Navy!"

By Capt. Edward J. Ruppelt USAFR

Although his early contributions to the field are typically revered for both his objectivity and his willingness to divulge military "secrets," Ruppelt abruptly changed his opinion regarding UFOs at some subsequent point, and whether or not the accusations that he caved in to official pressure to publicly retract his initial opinion are accurate, he did conclude finally that none of the phenomena were extra-terrestrial after all, and dismissed UFO reporters as unreliable.

According to his widow, he had become disillusioned by what he regarded as the questionable mental states of typical reporters of UFO-related events. Possibly this was simply an over-reaction to the ordinariness of the reporters, and as a no-nonsense military man with an educational background in engineering, Ruppelt was expecting too much of people if he expected everyone to be as calm and rational and scientifically-inclined as he was. But despite his seemingly unanimous endorsement in the field as "objective," if you read his books, it is possible to detect a distinct bias from the onset.

Although Ruppelt's investigations could hardly have been less thorough, he did use an official approach which separated reporters into distinct – and distinctly questionable – categories, one of which was labelled "CP," for "crackpots," and included everyone who claimed to have come into personal contact with an extra-terrestrial.

This represents a significant percentage of reporters, and if he dismissed them all summarily as "crazies," it is not surprising that he became disillusioned about the subject. And note that J. Allen Hynek, who enjoys an even greater status in the field, displayed a similar – if not quite so all-encompassing – "scientific" bias against reporters, dismissing as a crackpot anyone who claimed to have received a message of a religious nature. And this despite that fact that religion would seem to be single most common feature of ET-related messages.

APPENDIX K

Project Blue Beam

"Project Blue Beam" is the code name for a deep state government program alleged by investigative journalist Serge Monast (1945-1996). Monast, who is from Quebec, Canada, published a book of the same name ("Project Blue Beam," Presse Libre Nord-Américaine) in 1994, in which he exposed the plans for a decades-long deception perpetrated by a global cabal of financial elites, for the purpose of establishing a totalitarian global government. The goal, which is referred to as the "New World Order," was to dissolve all nation states in favor of a single, central control which would eliminate wars by eliminating religious/regional/political conflicts.[50]

After publishing his book, Monast was arrested for spreading false information. He died of a heart attack the day after his release, and his allegations have been vigorously debunked ever since. But while Monast's work, which is still in print and readily available online, ("Project Blue Beam," Ethos, 2023) is officially classified as a "conspiracy theory," it was taken seriously at the time, and has received renewed interest in the past several years because all but the final two of the pre-conditions which it outlines for the implementation of this alleged "new world order" have come true – and the final two are imminent. Moreover, his allegations cohere closely with

---

[50] This appendix was expanded into a longer paper which includes photos, entitled "Orbs, Drones, and the New World Order" (2025).

those made by establishment ufologist and disclosure advocate Stephen Greer, and at least two American presidents – George Bush Sr., and, more recently, Joe Biden – have explicitly used the very term "New World Order" to describe their goal for the future of the planet. [51] Others, like Barack Obama, have made cryptic remarks about how the elected government "isn't really in charge," and the general concept of "globalism" is now entrenched as not only desirable, but necessary for the very survival of our species.

Under Project Blue Beam, the prevailing global political order, which hinges on the concepts of individuality, national identity, competition, and religious freedom, would be replaced with a one-world, fascist government, and the disparate world religions, unified under a single,

---

[51]   The modern idea of a global government run by financial and political elites is thought to date back to 1776, when Adam Weishaupt, a German philosopher of law, is said to have proposed a such a scheme, creating a secret society called "The Illuminati" for the purpose. There is copious literature on the Illuminati both in print and online in various formats, but for an extensive summary of the "New World Order" concept in its current incarnation, see William Guy Carr's (politically incorrect) "Pawns in the Game" (Dauphin Publications, 2014).

Carr's explosive allegations, which have gained new popularity in recent years with the exposure of deep state programs like MK Ultra, mirrors that of "Project Blue Beam," purporting as it does to disclose he existence of a global deep state plan to seize control by a succession of steps which involve things like controlling the press, corrupting the youth, destroying traditional morality and values, and finally, various forms of terror and war, complete with underground bunkers and cities for the elite.

invented, "New Age" spirituality. [52]

The concept of religion is central to Project Blue Beam, and essentially what separates it from other total-control fascist political orders with global aspirations. Unlike communism, for example, The New World Order acknowledges the importance of religion to the masses, so instead of attempting to abolish it, not only creates a new religion, but elevates it to the status of "first principle." The idea is to appease believers internationally by creating a faith which combines elements of all of the existing world religions, while planting new archaeological evidence and implementing high-tech mind-control tricks like space-based holograms and machine-to-brain radio waves, to persuade non-believers of its veracity.

The second feature which differentiates the New World Order from other fascist regimes is that instead of *imposing* central control upon the masses, it creates conditions which are designed to make the citizens demand it themselves. In other words, terror techniques like manufactured natural disasters, pandemics, nuclear wars, and – in its final phase – a simulated alien invasion.

Any one of these things would seem to be sufficient to propose global governance of the world's population, but the idea with Blue Beam is to make the concept of global control *appealing,* and this takes both time, and advanced technology. The four aims of the project are as follows:

---

[52]   There is also a component which suggests that the god in this new world religion would be called the "Anti-Christ," but since the name of the entity is not critical to understanding the theory, its use will be avoided here.

1. Destroy religion.
2. Destroy nationalism.
3. Destroy individualism/individual creativity.
4. Substitute the state for family.

The project, which is alleged to have been engineered by NASA under the rubric of the UN, was to unfold gradually in a number of successive steps, beginning with military-induced earthquakes which would unveil new archaeological discoveries undermining essentially everything that we have been taught to believe about history, and culminating in the vindication of a New Age religion which would combine elements of all major world religions.

It is this new religion which is to serve as the basis of the new world order, and would be secured ultimately via the projection of giant holograms in space, featuring the prophets and gods of the three dominant world religions, all urging believers to accept that Islam, Judaism, and Christianity are, in reality, one.

This, incidentally, is already accepted by historians, who could presumably be counted on to verify at least the academic aspect of the merger. But even for devoutly religious people, it is not too great a leap – especially if aided by a holographic rapture featuring Jesus arriving to lead his flock triumphantly into the New World Order. Electromagnetic waves beamed directly into the brains of agnostics would be deployed to convince any remaining resisters, and, finally, a flat-out nuclear war to repel the alien invaders who would arrive at the same time to take over earth.

The specifics of how the New World Order is supposed to be accomplished must have sounded fantastic at the time that Monast published his book, requiring as they did sci-fi-worthy weapons and mind control technology, combined with political amorality of unprecedented proportions. But such weapons apparently now exist. And amorality has become quite commonplace in today's world, where traditional ethical systems appear to have been completely repealed, and replaced – at least in the political arena – with a "no-limits" utilitarianism where the end justifies any means.

When Monast exposed the plan, most of the preconditions for a new world order of some sort had already been achieved by a combination of evolving personal ethics and workplace changes deriving from social revolutions like the sexual revolution of the 1960s and the feminist revolution of the 1970's. The most amenable to the Blue Beam project, the collapse of traditional religious belief, was a direct consequence of all of the social upheaval which characterized these decades. And, given man's seemingly innate need for spiritual expression of some sort, the widespread erosion of traditional religion was followed quite naturally by the emergence of various forms of "cool" New Age religions – most notably those which featured extra-terrestrial components like psychic "channelers" from beyond.

The accompanying expansion of Marxist-based social sciences and interdisciplinary liberal arts programs in the universities, combined with media manipulation in the form of a government-controlled mainstream press and entertainment industry, had effectively eliminated all traditional values in Western political and educational systems, and the concept of "family" had been extended and re-defined so completely as to render it unintelligible.

All of these social changes – which were generally embraced as not only progressive, but positive – have, then, succeeded in establishing a socio-political environment ripe for the completion of a fascist global take-over. Add environmental crises, pandemics, and a sky full of strange objects, and all of the preconditions for takeover would seem to have been met; all that is wanting is a nuclear event.

If one examines post-Cold War history from this perspective, it all fits neatly within the rubric of an over-riding globalist project of some sort. Whether it is Blue Beam, or some other globalist plan, it is impossible not to notice that everything which has happened politically since at least 2020 seems "artificial," for lack of a better word. We are clearly being deceived about not just something, but about *everything*. By the middle of the Biden administration, there did not even appear to be an attempt on the part of the deep state to hide the fact that something was going on behind the scenes, and the elected government was not in control of anything.

Possibly this was part of the plan, but by the end of that administration, social engineering and the official dissemination of false information had become so blatant, and persecution, censorship, and even arrest of political detractors, so openly transparent, that the general public had become fearful, and running for political office on such a platform was literally life threatening, as several assassination attempts were made on the inadequately protected Republican opponent, and the leading Independent candidate was denied security protection, period.

As for the general public, it is no secret that a powerful faction of the New Left has trained what is now several generations of university students in new-Marxist globalist ideology, and the expansion of university education for the masses has enabled them to assert the will of their leaders democratically in every aspect of society.

Much of this, apparently as per the plan, consists in destroying history, national pride, individualism, traditional religion, and the family. But note that the globalist movement associated with the New Left is explicitly political; it does not address any of the central components outlined in Blue Beam. In fact, the present New World pre-Order, so to speak, espouses an emotion-driven, feminist agenda which is explicitly anti-science, and, when it is not disparaging all things military and industrial, is simply mute about technology, space, and extra-terrestrials. Moreover, the present alien invasion simply does not look like a false flag.

At the time of this writing, the sky is filled with unidentifiable objects of some sort, all of which have the appearance of classic extra-terrestrial craft. Some of them look like metallic orbs with plasma-fuelled propulsion systems, and others, like Tic Tacs, transparent portals, and the cosmic "wheels" described in the Bible as "angels." Others are huge, army-issue type drones of unknown origin, hovering over military installations in a defensive fashion, or "chasing" the UAPs in the upper atmosphere. None of this has the slick, planned, appearance of a false flag, and the elected government, who would have to be in on such a plan, does not appear to have any idea what is going on.

The current space crisis also lacks the simple and obvious, pre-planned "solution" and ulterior motive which comes with a false flag operation. On the contrary; there is no obvious agenda, let alone an ultimate goal, to any of this. The entire problem appears to be as insoluble as it is inexplicable, and nobody in charge seems to know what to do, or what to expect next. But most telling, the solution implemented to solve the problem is not working.

The drones, which have the appearance of being deployed

to protect sensitive military installations and critical infrastructure from extra-terrestrial attack, are not repelling anything, but rather being themselves destroyed by space invaders which do not look "fake," and cannot possibly have been created and unleashed globally by any co-ordinated terrestrial power. Even the propaganda press is admitting that we do not know what any of these things are.

If this is a false flag, then the objects in space are holograms. But the drones are real, and holograms cannot destroy real objects. Even the fact that the "holograms" are outrunning, eluding, and escaping the drones which have been unleashed to defeat them, suggests that this is not some over-sized, space-based stage show, put on for the purpose of achieving a pre-planned political end. If one compares the present alleged false flag alien invasion to other alleged false flag events, the difference is stark.

The typical false flag operation is simple, swift, and dramatic. It is comprised of a well-defined goal, a clear, easily identifiable plan, an obvious scapegoat, and a complicit, explanation-ready press. It also seems a bit "off," somehow. The event is too convenient, too out-of-the-ordinary, and the explanation, too easy. There are too many details surrounding the entire event which just don't "fit."

Sometimes, such as in (alleged) false flag terror attacks like 9-11, whistle blowers, technical experts, and first responders speak out immediately, and conspiracy theorists submit their debunks within minutes, pointing out every inconsistency in the official explanation, and providing credible hypotheses as to what the goal of the covert operation is. In other cases, such as Pearl Harbor, the official explanation prevails for decades before the truth finally emerges. But when it does, it makes more sense than the lie.

Yes, of course! President Roosevelt wanted badly to participate in the war in Europe – so badly that he colluded with the Japanese, and leaked information pertaining to a (suspiciously unguarded) military outpost in Hawaii. One attack was all that it took to convince the entire country to enlist, and both Congress and the able-bodied men who were required to be sacrificed stepped up immediately. The lone officer who was blamed for the attack was just another useful pawn in the game, and the ultimate goal – the defeat of fascism in Europe – was sufficient justification for all of these things and more.

Whatever one thinks of this revision of history, it all fits, whereas the official explanation doesn't quite add up. This is the mark of a false flag, and it doesn't look anything like the present UAP invasion, despite what both Blue Beam and disclosure advocate Stephen Greer insist.

In the present alien invasion, there is no apparent goal, let alone any official explanation. The officials quite clearly do not know what is happening, and have stated as much. There is no single, dramatic, planned-appearing strike or significant event. This invasion is taking place in the skies globally, and appears almost haphazard, rather than carefully planned and co-ordinated. And it certainly is not publicized and presented as a dire, immediate problem with a pat solution.

The press has been instructed to hush everything up, and even the conspiracy set can offer no credible hypotheses to explain what is going on in the skies. The few whistle-blowers, most of whom are in the military, state simply that "There are aliens," or, more ominously, say "Get your affairs in order, stockpile enough survival goods for six months, and pray to your god."

The UAP objects in the sky have been arriving and

amassing gradually for at least months, and arguably for years, and there is neither a distinct "problem," nor any solution at all, let alone an obvious one. There is also no apparent goal to the operation – if it is an operation.

If it *is* a planned, false flag alien invasion, with the goal of terrifying the population, then it has been a dismal failure; the general population appears to be enchanted by the extra-terrestrial presence in the skies, and people all over the world are out in full force, taking videos with their cell phones, and calling out "Welcome, alien friends!"

The military-issue drones which have been deployed in an apparent attempt to protect our nuclear installations, on the other hand, are reviled as sinister artifacts being cranked out by our evil terrestrial governments, whom people are openly inviting the aliens to overthrow, and save earth from their obviously immoral, if secret, deep state agenda.

But regardless of whether or not the current "space event" is a failed Blue Beam alien attack designed to terrify the public, the conceptual goals of Project Blue Beam have been accomplished. The first phase, which involves weather manipulation and the undermining of historical beliefs about religion, is also well underway, as the dictates of government-approved climate science have become the new morality, and UFO disclosure advocates posit an identity between real-life extra-terrestrials and the "supernatural" entities featured in holy books like the Bible.

The most prominent of these pro-UFO advocates have gone so far as to invent a new, UFO-based system of spirituality which incorporates the essential elements of Eastern religions, and combines them with the alleged existence of deep-state-suppressed technology, and energy sources which would fuel the entire globe, completely free of cost, and eliminate war.

We have also not only experienced the predicted global pandemic, but have been informed that the protocols for dealing with it were already in place when it broke out. But the important thing to note is that the technological advances required for the completion of the project are also now available: Military scientists have succeeding in creating machine-to-mind communication, sophisticated, space-based hologram projection, and, most critically, authentic-looking UFO-replica spacecraft with which to stage a false flag alien invasion, should they ever wish to.

They have also allegedly succeeded in not only designing weather control "weapons," but are actively creating widespread extreme climate events like earthquakes, and reporting daily discoveries of new archaeological evidence to discredit everything that we had previously been taught about both geology and anthropology, while classics scholars are re-interpreting ancient texts in a literal, non-mythological, way, and alternative anthropologists, insisting that the other-worldly creatures depicted in ancient artifacts are not creative imaginings, but real entities, who came to earth from outer space, bringing both their knowledge, and their religion, with them.

New varieties of intelligent species seem to be coming out of the woodwork all at once, as a colorful assortment of never-before-known hominids of all shapes and sizes are reported in the news, alongside discoveries which confirm the existence of Biblical figures and artifacts like Noah's arc, and verify the birthplace of Jesus.

Non-believers are being urged to accept the empirical reality of Biblical events, in other words, and urged to embrace the book's so-called "mythological" elements as extra-terrerestrial. The government has also gone on record as confirming the existence of an extra-terrestrial presence on earth today, and promises a full reveal soon.

Also in apparent accordance with the Blue Beam agenda, whistleblowers like Edward Snowden have revealed that the government is secretly compiling data about every citizen, and not only eavesdropping on our conversations, but recording our every purchase, image, and movement – facilitating, it is alleged, unique, machine-to-mind messages personally tailored to turn every non-believer into a devout, New Age supplicant.

The religious transformation of society, then, is already well underway, regardless of the catalyst behind it. And note that it isn't necessary to believe that the gradual turning away from traditional religious beliefs and morality was planned and directly caused by the architects of Blue Beam, to believe in the existence of the program; it is far more likely that Blue Beam was devised as a response to evolving beliefs about these things, and simply capitalized upon them.

The phasing out of cash currency, which is supposed to aid in removing individual autonomy, is also almost complete. And, exactly as per Blue Beam, it has not been necessary to "trick" the population into accepting any of these things. They were not only accepted without protest, but positively embraced by the masses, for the ease with which they permitted people to go about their daily lives, free from the strictures of traditional religion, and the necessity of being forced to visit banks in order to make financial transactions, or sit at home waiting for phone calls.

The accompanying loss of privacy has been accepted almost without murmur, as citizens demand more and more conveniences associated with life online. Many people now not only do all of their shopping and banking in this way, but renew passports and prescriptions, and even attend the bulk of their medical appointments remotely, chafing at the prospect of appearing *anywhere* in person.

The pandemic extended this practice still further, with out-of-the-office employment and education opportunities, all of which created new ways for our governments to surveil and collect a mountain of minutiae about its citizens – culminating in the collection of individual DNA samples extracted while injecting government-mandated, military-issued COVID vaccines.

What they intend to do with all of this information is anyone's guess, but according to believers in Project Blue Beam, it is all related to the development of personalized, universal mind-control by a central global power, which will render us all unwittingly its puppets. Its precursor, AI, has not only arrived, but has been enthusiastically embraced by everyone, effectively eliminating the need for individual human creativity.

Whether this theory is true is impossible to confirm, but what can be confirmed is the existence of a number of legal "loopholes," implemented by the Biden administration, which would, if utilized, facilitate the loss of national sovereignty requisite to complete global takeover.

The first of these permits the World Health Organization – a secretive, highly undemocratic global entity – to assume control of America in times of emergency, where the term "emergency" is not defined. In other words, it is up to the WHO, and not Congress, to decide what does and does not constitute an "emergency," and it does not have to be a health-related crisis like a pandemic.

Another little-known legal channel for global take-over is related to central control of information dissemination, and it involves classifying independent journalists as "domestic terrorists." Cyber currencies fall under this same rubric, the allegation being that, as independent, citizen-managed assets, they can be used by gangsters for illegal – and untraceable – financial exchanges. Paper money has this same feature, of course, and this is precisely why it is allegedly being phased out in favor of electronic transactions which can be closely monitored. The online-only aspect of the single electronic currency outlined in Project Blue Beam has already been essentially achieved, and its purpose – the ultimate control over every citizen – has been tested. It works.

Online banking has provided governments with an especially effective means of controlling its citizens directly, as it enables them to freeze or seize bank accounts in order to enforce compliance. This has already been done in Canada, where, towards the end of the pandemic, a convoy of truckers converged on the capital, overtaking the city center with huge, live-in rigs which they refused to remove until the government repealed its vaccine mandates.

The truckers, who both outnumbered, and ignored the police when ordered to leave, were supported by the public internationally, as well as the army, and set up camp with their children, while local citizens brought them food and joined their parties. When told to vacate the space, they barricaded themselves in their trucks instead, honking their horns continuously, or gave interviews to journalists, boasting that they lived in a free country, and had the right to protest. And then just when everyone assumed that citizen's rights had prevailed, and expected some government representative or other to address the crowd to negotiate terms, the truckers were defeated instantly when the government froze their bank accounts, making it impossible for them to buy gas or supplies. The incident has been repeated in horror in the indie news ever since, as a chilling reminder of how our governments have become agents of control, rather than representatives of the people.

Regardless of whether this is an accurate interpretation of that particular event – which is a matter of some contention, with many viewing the trucker takeover as a genuine insurrection – it does serve to illustrate the ease with which governments have acquired the ability to exercise direct control over citizens in this internet age, as well as to regulate the flow of information. It also serves as an excellent example of how personal, and even financial autonomy has been willingly relinquished by citizens, in favor of technology. And again, the above points do not in any way establish the truth of Project Blue Beam, but they are consistent with such a project, and are certainly conducive to its end, if the project does exist.

The final aspect of Project Blue Beam under consideration is also the most chilling, as it involves the use of nuclear weapons – which, if only co-incidentally, is now being seriously threatened on an almost daily basis. The idea here is twofold: First, destroy vast swaths of the planet and kill untold numbers of people with nuclear bombs. Then, when the destruction is complete, admit to the survivors that the entire "war" was planned, and expect them to be so horrified, and so appalled at the real-life consequences of these "deterrence" weapons, that they will unite in demanding that all nuclear weapons be destroyed immediately – and, more importantly, that some central authority exercise strict control of the global population to prevent something like this from ever happening again.

Project Blue Beam goes on to envision a world where there is no human independence or personal autonomy, and it is possible to view videos online which depict robotic citizens of this New World Order, who live in what looks like a cross between a high security psychiatric institution and a toxic-carpeted mid-Soviet Union apartment complex, and never leave their homes, but have all of their (state-mandated and strictly rationed) requirements delivered to their doors by world-state Fed-Ex men in sterilized space age jumpsuits. It all looks preposterous.

But the important thing to remember is that the historical rendering of such a plan is completely separate from its contemporary YouTube imaginings. And if the plan sounds as absurdly dated as it is preposterous, it should be noted that it isn't necessary for the current deep state cabal which would be responsible for implementing the plan, to follow it to the letter; it can always be edited, updated, or abandoned as unfeasible at any point along the way. And regardless of the program's status relating to any of these things, if one examines the general overview for accomplishing globalism outlined in Blue Beam, it is actually quite ingenious – especially with respect to the idea of creating a situation so intolerable that people positively *demand* to be placed under some central authority. In fact, we are in exactly such a situation now.

Only a negligible minority of the population is doing the demanding, of course. But it must be pointed out that they do seem to be the ones with all of the power, and all of the legal requirements for accomplishing global government are already in place. It remains to be seen whether the billions of nation-state citizens worldwide who are railing against globalism would change their minds in the face of an alien invasion and/or nuclear war, but as a theory, the existence of a plan such as Blue Beam does make sense of a bewildering variety of otherwise-inexplicable present socio-political events which have clearly been planned by *someone,* for *some* purpose.

It accounts for the dramatic expansion of non-classical liberal arts programs in universities, all of which are undisguisedly Marxist in orientation, and not only indoctrinate rather than educate, but actively encourage students to despise their country, disparage its history, and even destroy its artifacts.

It explains the emergence, and finally the entrenchment of race-based political policies, the government-approved classification of citizens as ethnic collectives rather than individuals, and the emergence of non-competitive entrance requirements and grading schemes.

It accounts for the proliferation of mysterious underground bunkers designed to withstand extinction-level catastrophes.

It explains the obviously pan-national new conventions, rules, and protocols imposed upon all UN member governments during and after the pandemic – all of which were transparently calculated to destroy traditional conceptions of human families and even biology.

It explains the persecution and arrest of political opponents who have "outsider" status, and the united front against them put on by both mainstream Republicans and Democrats.

It explains the government takeover of the press, control of social media, the crackdown on free speech, and the drastic steps taken to control the public beliefs a via "fact checking" Department of Truth.

It explains the relentless collection of seemingly insignificant information about every aspect of every citizen's life.

And finally, it explains the abrupt reversal of the policy of keeping all matters extra-terrestrial top secret, and official line that they do not exist, for an official admission of the extra-terrestrial presence on earth, accompanied by the insinuation that they pose a serious threat to our national security.

What Blue Beam *doesn't* explain is *why*. Why, that is, are these secret globalist elites so intent upon exerting ironclad control over the entire human population, if all they want is to dispense with our nuclear capability? They could do that without our consent. But in fact, they *have* our consent. It is governments, and not citizens who are so fond of these weapons. Citizens routinely protest against them, and nobody wants either nuclear weapons, or conventional wars.

Generally speaking, when a government wants to control people, its aim is to uphold and/or extend a political ideology – which, incidentally, its enforcers believe in. But the problem with conspiracy theories is that this is the reason given for everything: The deep state wants to *control* people. The concept is so pervasive, and the statement so obviously true, that nobody seems to point out that control for its own sake makes very little sense. In the case of the Blue Beam conspiracy, *nothing* makes any sense. Unless, that is, the agency wanting to control humans is not some shadowy secret global government, which craves power for its own sake, and hobnobs with extra-terrestrials for the purpose of acquiring their advanced weapons technology, but the extra-terrestrials themselves.

Consider the facts: We know that the primary reason why extra-terrestrials have come to earth is to dismantle our nuclear capability. For some reason, they do not seem to be able to use their advanced technology and alleged supernatural powers to do so directly, but have been attempting (unsuccessfully) to convince humans to do this for them – or, as they put it, "for ourselves."

Their primary technique for doing this consists in an odd – and oddly uncomprehending – appeal to religion, which is also the cornerstone of Blue Beam.

Their willingness to nuke a large percentage of us in order to achieve this end has the same chilling, amoral character that the extra-terrestrial "abduct-and-hybridize-humans" scheme does; it is exactly as though the architects of Blue Beam do not regard humans as sufficiently sentient to qualify as intrinsically valuable, moral agents.

The "eliminate individuality" aspect of the plan is similarly anti-human, and, like the idea of appealing to religion as a means of control, oddly uncomprehending. In fact, the concept of total control is itself distinctly alien in nature, as is the collection and retention of this seemingly pointless mountain of information about us. It simply does not make any sense – unless they intend to use us for some purpose of their own.

www.ingramcontent.com/pod-product-compliance
Lightning Source LLC
Chambersburg PA
CBHW072144230526
45467CB00040B/47